POPULAR MECHANICS

THE ART OF
MECHANICAL DRAWING

Popular Mechanics

THE ART OF MECHANICAL DRAWING

A PRACTICAL COURSE FOR DRAFTING AND DESIGN

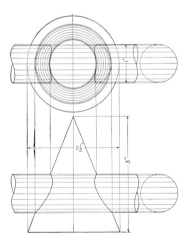

WILLIAM F. WILLARD

HEARST BOOKS

A division of Sterling Publishing Co., Inc.

New York / London
www.sterlingpublishing.com

Library of Congress Cataloging-in-Publication Data

Willard, William Franklin, 1879-
  Popular mechanics : art of mechanical drawing/William F. Willard.
    p. cm.
  Originally published: Practical course in mechanical drawing for individual study and shop classes, trade and high schools / William F. Willard. Chicago, Popular Mechanics Co., 1912.
  Includes index.
  ISBN 978-1-58816-759-0
  1. Mechanical drawing. I. Willard, William Franklin, 1879- Practical course in mechanical drawing for individual study and shop classes, trade and high schools. II. Popular mechanics magazine. III. Title. IV. Title: Art of mechanical drawing.
  T353.W65 2008
  604.2–dc22

                                        2008007273

Book design by Barbara Balch

10 9 8 7 6 5 4 3 2 1

Published by Hearst Books, A Division of Sterling Publishing Co., Inc.
387 Park Avenue South, New York, NY 10016

Popular Mechanics and Hearst Books are trademarks of
Hearst Communications, Inc.

www.popularmechanics.com

For information about custom editions, special sales, premium and corporate purchases, please contact Sterling Special Sales Department at 800-805-5489 or specialsales@sterlingpublishing.com.

Distributed in Canada by Sterling Publishing
c/o Canadian Manda Group, 165 Dufferin Street
Toronto, Ontario, Canada M6K 3H6

Distributed in Australia by Capricorn Link (Australia) Pty. Ltd.
P.O. Box 704, Windsor, NSW 2756 Australia

Printed in China

ISBN 978-158816-759-0

CONTENTS

FOREWORD

The turn-of-the-last-century Renaissance man knew how to do everything, from catching dinner to building the table he'd eat it on. One hundred years later, many life-altering innovations have changed our daily lives; things move faster, more efficiently, and with greater precision. . . right?

With the introduction of each new technological advance in our lives, whether it's a calculator eliminating the need for long division or email replacing handwritten letters, we lose something. Slowly but surely, skills once considered essential are dismissed as unnecessary and archaic.

Nowhere is this truer than with mechanical drawing. As a result of breakthrough computer-aided drafting and design software, hand-drafting is becoming redundant. Budding engineers and professionals now learn to draft on a computer and can design a dizzying number of plans in record time—usually with amazing accuracy.

But even the finest computer program, especially software dependant on user input, makes mistakes. Drafting by hand, while a much slower process than drawing with whiz-bang software, can still be valuable, even today. Armed with an understanding of drafting, you would no longer have to rely solely on the computer for accuracy and would be able to verify blueprints and designs for yourself.

Learn, develop, and practice the classic art of hand-drafting with this step-by-step instruction manual, originally published by Popular Mechanics in 1912, and follow the example set by the turn-of-the-century draftsmen who first learned from its pages. Starting out on paper and vellum gives you a renewed appreciation for the tactile pleasure of mechanical drawing—and a powerful awareness of the steps today's software takes to produce results.

This fully functional guidebook to the ins and outs of hand-drafting has been updated slightly to reflect today's high-quality tools and materials. We've changed nothing else, and the style of writing still occasionally echoes the tenor of the simpler time when it was first printed. We're sure you'll find that the beautifully intricate illustrations and precise exercises are as useful now as they were first published.

The days of the Renaissance man are perhaps over. But you will still find satisfaction and value in striving to learn as much as he knew. Enjoy this essential guide-book to a lost art, courtesy of Popular Mechanics.

The Editors
Popular Mechanics

CHAPTER I

INTRODUCTION TO PRACTICAL MECHANICAL DRAWING

Mechanical drawing is one of the most popular and most profitable subjects of study today. It is an essential qualification for most areas of engineering and an almost indispensable accomplishment for many other occupations—often the secret of successful advancement. It is founded upon the branch of mathematics known as geometry which, as applied to drawing, becomes a delightful and interesting subject—and not the difficult abstraction the beginner fears. Suppose a farmer who already knows the diameter and height of a tank needs to find out how many gallons of water will be needed to fill it—or how many bushels of wheat will fill a bin, or how many acres there are in a field a quarter of a mile square. These examples illustrate the practical application of geometry, a subject no less important to the mechanic than the farmer, but a thousand times more interesting to the student than those exercises found in the usual textbook.

In preparing this manual the author was mindful of the many circumstances and limitations that have so often combined to deny to aspiring young people the advantages of a complete education. In this day and age competition and industry conditions demand the best training and skill for every productive effort. What artisans or mechanics do to improve intellectually therefore increases their efficiency and value to their employers in every respect.

In geometry a student is concerned with the theorem behind a problem, and its proof, or why it is so. In mechanical drawing, the mechanical operations of construction, the actual mapping out of a problem graphically with the use of compass, triangles, and other instruments, is considered essential. However, this course does not preclude a mastery of the principles of geometry. Anyone who is aware can apply, to a good advantage, the geometry exercises that follow to some project or other in order to invent something or puzzle it out, even without first having studied the subject.

Surveyors with their tapes and transits, and architects or mechanical engineers with their slide rules and formulas, must also know these exercises. If a craftsman desires a brace or bracket for a plate rail, he must know how to lay out the desired curves and angles. Similarly, before making a taboret or jardinière stand with a hexagonal or octagonal top, the craftsman must first solve the geometric problem—or be unhappy with the results.

One reason for not accepting a freehand perspective sketch as a substitute for a geometric drawing is that the sketch seldom shows all the information the workman requires. The sketch deals with outward appearances only, and from just one viewpoint. The mechanical drawing of an object delineates the actual facts, inside and out, and from as many viewpoints as the object has dimensions. Any hidden or detailed information is considered as important as that which is visible, and these details are represented by suitable conventions (the word convention, in drafting, meaning a customary symbol or method established by precedent).

A freehand sketch is governed by well-known laws of perspective that constitute the language of the artist from an aesthetic standpoint. A mechanical drawing is represented by customary shop and drafting-room conventions and is the language of the mechanic and artisan. The former develops the power of observation, good judgment, and individuality; the latter, precision, accuracy, and mechanical ability.

The mechanical drawing has an advantage over the sketch: a workman will not be apt to confuse apparent dimensions, as seen in perspective, with true measurements, as seen in workmen's drawing. All working drawings are made to scale, and all dimensions are proportional and properly placed. Such a drawing must be made with all those who must use it in mind, and in such a manner that the least skilled person in the shop will understand them. Otherwise, any errors in construction will be blamed on the draftsman.

Mechanics, designers, engineers, and artisans of any trade fully realize the importance of a definite plan of procedure. Bridges, buildings, railroads, and canals must be planned out on paper, and their feasibility confirmed, before a mechanic or construction company begins the actual work. Perhaps the most important part, if not the most difficult, is making the plans and specifications. The next most important part is working according to the plans' directions.

Constructive drawing also finds expression in a multitude of shops. A cabinetmaker, machinist, patternmaker, or contractor must have intelligible pictures or drawings to guide his hands, and these drawings must be accurate and clear. A draftsman, whether amateur or professional, who fails to make them so may, through ignorance, carelessness, or both, cause a loss of great consequence to his employer and the world at large. Someone has said: "Mechanical drawing is the alphabet of the engineer; without it he is only a hand. With it he indicates the possession of a head." Indeed, the hand will only do what the head directs.

A uniform code of conventions and symbols is required among workmen and shops. Such a language, if it may be so called, has come to be accepted among draftsmen, who adhere closely to the modern approved forms; these will be used throughout this manual.

[CHAPTER II]

THE DRAFTSMAN'S EQUIPMENT

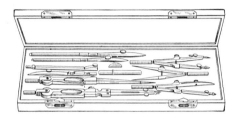

The old saying that a poor workman blames his tools is almost always true, for if a workman is content to work with an instrument poor in quality or poorly kept, it can be inferred that he expects to do poor work. How can a draftsman produce an accurate drawing with a nicked T-square, warped triangles, a dull pencil, a ruling pen clogged with ink, a compass with wobbly legs on blunt points—all imperfections that would mar the finished drawing? To do commendable work one must use high-quality materials, take excellent care of them, and keep them in good repair. A complete list follows:

1. *Drawing board:* It should be at least 18" x 24", and preferably 23 x 31" (Fig. 1, page 16). Some are made of kiln-dried wood and have a metal edge; other, higher-end models are made of Melamine. Some come equipped with clips to hold drawing paper, a built-in straightedge or T-square, or small, foldaway legs to that prop them up at an angle.

2. *T-square:* It should be as wide as the drawing board and made of aluminum (Fig. 1).

3. *Drawing paper:* Always use acid-free paper. Both Strathmore and Utrecth make serviceable pads of white two-ply bristol in 14" x 17" and 19" x 24".

4. *Artist tape:* Make sure it is pH neutral.

5. *Staedtler pigment liner four-pen set* (contains 0.1, 0.3, 0.5, and 0.7mm waterproof drawing pens). A more expensive alternative is the Rapidograph made by Koh-I-Noor, which comes in a wider variety of sizes.

6. *Pencils and sharpener:* Staedtler 2H and 4H. Sharpen with an Alvin rotary lead pointer so that the point is exposed $\frac{5}{16}$" or $\frac{3}{8}$".

7. *Spring-bow compass with double-hinged legs, a lengthening bar, and a universal adapter* that fits blades and writing instruments. Inexpensive compass sets are also available.

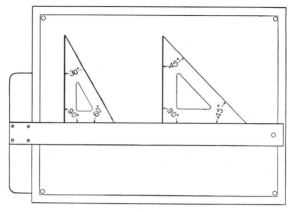

FIG. 1

8. *Box of leads*

9. *Erasers:* A Staedtler Mars white plastic eraser and a Design art gum eraser for pencil, Imbibed eraser strips for ink, or others of similar quality.

10. *Erasing shield*

11. *Rulers:* An ordinary ruler will do at first; however, an architect's or engineer's tri-scale, which is divided into different measuring systems, is best. It should be at least 12" long.

12. *Triangles:* While available in aluminum, plastic triangles are easier to work with because they are transparent. Two will be necessary: a 30°-60°-90°, available in 8" and 10" lengths; and a 45°-90°, available in 6" or 8" lengths (Fig. 1).

13. *Protractor*

14. *French curves:* Plastic curves are usually available in sets of four or eight.

The equipment need not be expensive; if it is convenient, consult with a working draftsman before selecting the materials.

HOW TO USE THE MATERIALS

1. Always use the upper edge of the T-square, which should be held against the left edge of the drawing board. Never cut against the upper edge of the T-square: the least nick will cause later work to be inaccurate.

2. The ruling pen should incline slightly in the direction of the line being drawn and be held so that the nib does not touch the edge of the T-square. Lines are ruled from left to right and from the bottom upward.

3. To get clean, sharp lines, sharpen pencils frequently. Never make ridges on the drawing by pressing heavily on the pencil.

4. Compass legs are jointed so that the lead can be square to the surface of the paper while a circle is being drawn. The hand should describe the

circle above the paper while holding the compass, not remain stationary.

5. Use a soft gum eraser to clean the drawing before inking in order to retain the black lines' glossy finish. Use ink erasers for pencil lines only in exceptional cases—if the pencil has caused deep ridges in the paper, for example. All division lines should be erased before inking.

6. All dimensions should be stepped off from the scale or ruler, with dividers, and then pricked lightly in the required place on the drawing. Instructions for how to scale a drawing properly will be provided later.

7. The triangles, which are used to draw oblique and perpendicular lines, should rest upon the upper edge of the T-square. Many and various combinations of angles may easily be made by combining both the 30°-60° and 45°-90° triangles. The oblique side of the triangle should always be to the right while in use, whether inking or penciling (Fig. 1).

8. French curves are used to define curves impossible to obtain with the compass. Although they are composed of many curves, they seldom have the right one, so it is often necessary to

shift it into many positions before the required results can be obtained.

9. The protractor is a semicircular instrument graduated into 180°. It is used to measure angles other than those obtained with the triangles.

10. These instruments must be kept clean. Use soap and water and a soft cloth or, for ruling pens, a Q-tip. Brand-name cleaning solutions and kits are also available for ruling pens.

CHAPTER III

GEOMETRIC EXERCISES WITH INSTRUMENTS

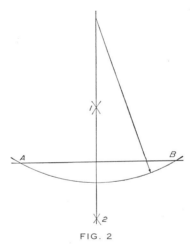

FIG. 2

EXERCISE 1

Bisect a line of any length and an arc of a suit-able radius.

Construction: With radius greater than ½ of AB and points A and B as centers, describe inter-secting arcs at 1 and 2. If a line is drawn from 1 to 2, it will bisect AB (Fig. 2).

EXERCISE 2

Erect a perpendicular to a given line from Exercise 1.

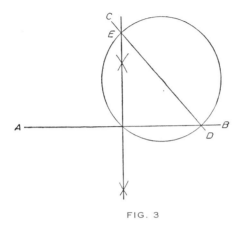

FIG. 3

SECOND METHOD:

Construction: From a given point E outside the given line AB, draw a line at any angle to AB. Bisect and inscribe a circle about ED. The circle cuts AB at a point of the perpendicular through point E (Figs. 2 and 3).

EXERCISE 3

Draw a parallel line through a given point X to a given line AB.

Construction: From any point B on the given line and a radius equal to BX, describe the arc. From X and the same radius, describe a similar arc through B. Mark off

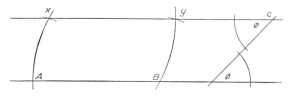

FIG. 4

BY on a second arc equal to AX. A line drawn through X and Y is parallel to AB (Fig. 4).

SECOND METHOD:

Construction: Draw a line making any angle with AB. With C as the center and any radius, describe an arc making angle Ø. Duplicate this angle with the given point as its center (Fig. 4).

EXERCISE 4

Divide two lines into proportional parts.

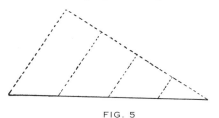

FIG. 5

Construction: Mark off one line into any number of divisions. Connect the extremities of each line. Using triangles, draw parallels through the remaining points (Fig. 5).

EXERCISE 5

Construct tangents to a given arc of any radius.

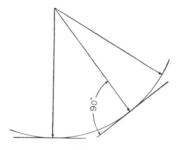

FIG. 6

Construction: With any radius describe the arc of a circle. From the center of the arc to any point of the circumference, draw a radial line. At the extremity of the radial on the circumference, erect a perpendicular. This is the required tangent (Fig. 6).

EXERCISE 6

Duplicate and bisect a given angle.

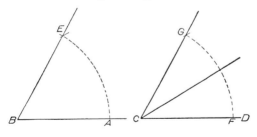

FIG. 7

Construction: Draw any two intersecting lines, making any convenient angle. To duplicate, draw CD with any length. Describe an arc cutting the given angle at A and E. With the same radius, describe an arc cutting CD at F. Mark off, with F as the center, the distance AE, and draw the other side of the angle through C and G. Bisect as in Exercise 1 (Fig. 7).

EXERCISE 7

Draw a 60° angle.

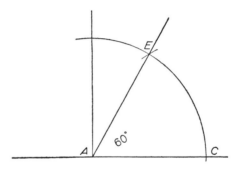

FIG. 8

Construction: With any line as its base and any point therein as its center, describe an arc of any convenient radius, cutting the baseline at C. With C as a center and radius AC, describe an arc at E. A line through AE is 60° to AB. Bisect it to get an angle of 30°. Other angles can be easily determined. (Fig. 8).

Note: 22° 30' reads 22 degrees and 30 minutes or 22$\frac{1}{2}$ degrees.

> 60" (seconds) = 1 minute (')
> 60' (minutes) = 1 degree (°)
> 360° (degrees) = 1 circle

The characters ' and ", which are used to designate minutes and seconds, are also used also to designate feet and inches. The context will, however, generally help avoid confusion as to their meaning.

TRISECTING ANGLES

EXERCISE 8

Using only triangles, divide a semicircle into angles of 15°. Use a T-square as a base for the triangles (Fig. 9).

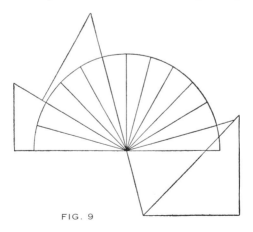

FIG. 9

EXERCISE 9

Rectify a quadrant of a circle. Approximate methods.

Construction: Draw a circle of any suitable diameter and divide it into quadrants. Draw a tangent of indefinite length at the lower end of CD. Through A draw a line at

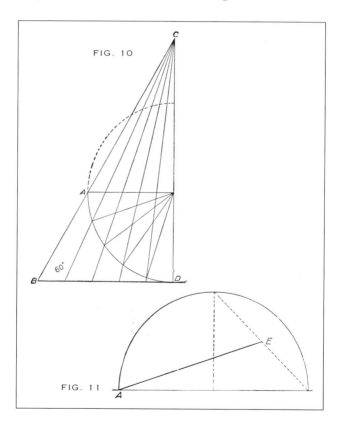

FIG. 10

FIG. 11

60° to this tangent. Where it cuts BD, determines the length of the arc AD. Any smaller arc can be determined by extending, through C and the other end of the given arc, a line to BD (Fig. 10).

SECOND METHOD:
Approximate (Fig. 11). AE = BD, (Fig. 10).

EXERCISE 10

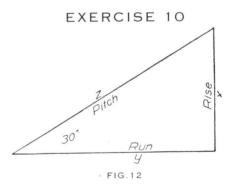

· FIG.12

Construct a right triangle one angle of which is 30°. The sum of all angles of any triangle is 180°. If a right angle is 90°, what must the remaining angles be? This exercise can be applied to determining the pitch or length of a rafter when its rise and run are given. The ancient Greek mathematician Pythagoras discovered that the square of the rise plus the square of the run equals the pitch squared: $x^2 + y^2 = z^2$ (Fig. 12).

EXERCISE 11

Approximate the distance across an unknown area by means of similar right angles.

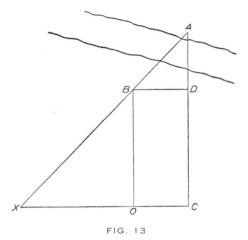

FIG. 13

Construction: Select a tree or object on the opposite side of the street, as indicated by A. Select another on this side (D). Mark off a convenient distance from D to C in the line ADC. Select a point B at right angles to AD and construct a parallelogram BODC. Determine point X on the ground that is in line with OC and BA, and measure the distance XO. By proportion:

AD : BD :: BO : XO; ∴ AD = BD x BO/XO.

Lay out the diagram, substituting known values for BD and DC, and solve the equation (Fig. 13).

EXERCISE 12

An equilateral triangle. *Construction:* Assume any length for a base. With a radius equal to the length of the base and each terminal A and B as centers, describe an intersection at X. Connect this point by lines to C and D. Measure

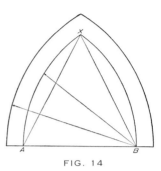

FIG. 14

the angles of an equilateral triangle in degrees. What is their sum? Stained glass windows are often laid out in Gothic arch forms using this kind of triangle (Fig. 14).

EXERCISE 13

FIG. 15

Isosceles triangles. *Construction:* On a line of given or assumed lengths and with a radius greater or smaller than AB, proceed as in the problem above. Are all the angles equal? What is their sum? (Fig. 15.)

EXERCISE 14

The vertex angle of an isosceles triangle is 150°, and its base is 3" long. Without a protractor, make a drawing. The trillium is an early spring flower shaped somewhat like an isosceles triangle.

EXERCISE 15

A scalene triangle.
Draw a triangle whose base is $2\frac{1}{2}$" and base angles are $22\frac{1}{2}°$ and $37\frac{1}{2}°$. What is the sum of the angles? What is the sum of the angles of any triangle? (Fig. 16.)

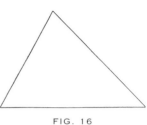

FIG. 16

EXERCISE 16

A circle within a square within a circle.
Inscribe a square within a 3" circle. Now inscribe a circle within the smaller square (Fig. 17).

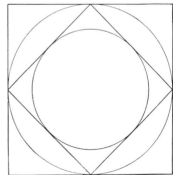

FIG. 17

EXERCISE 17

Draw a square whose sides are tangent to the larger circle in Exercise 16. What is the relation of inner to outer square? The syringa, or mock orange, is a flower with four petals in the shape of a square.

EXERCISE 18

A pentagon within a circle.

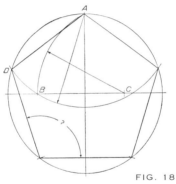

FIG. 18

Construction: Bisect the diameter of a circle. Bisect a radius. With C as a center and AC as a radius, describe an arc at B. With A as a center and AB as a radius, describe an arc on the given circle at D. AD is the length of one side of the polygon. Mark off the remaining sides and draw a five-sided star (Fig. 18).

What is the size of an interior angle? (Use a protractor to measure it.)

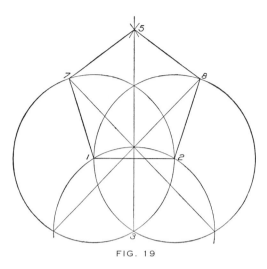

FIG. 19

EXERCISE 19

A pentagon.

Construction: Draw a $1\frac{1}{4}$" base. With one radius equal to the base length, describe arcs from centers 1, 2, and 3. Connect 1 and 2 with 7 and 8, respectively, and complete the pentagon. Inscribe a circle within the figure. Circumscribe a circle about the polygon. Many flowers, including pansies and violets, are pentagonal (Fig. 19).

EXERCISE 20

A hexagon within a circle.

Construction: Mark off the radius six times on the circumference of a circle and connect the points. Without using a protractor, determine the interior angle of this polygon. Use this formula:

$$2n - 4 \text{ right angles}$$

when n equals the number of sides of the polygon.

$$\frac{[(2 \times 6) - 4] \times 90°}{n} = 120°$$

Prove this to be true. Draw a six-pointed star (Fig. 20).

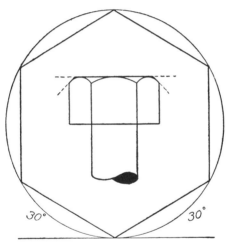

FIG. 20

EXERCISE 21

Draw a hexagon using the 30°-60° triangle.
The hexagonal bolt is an illustration of the use of the hexagon (Fig. 20).

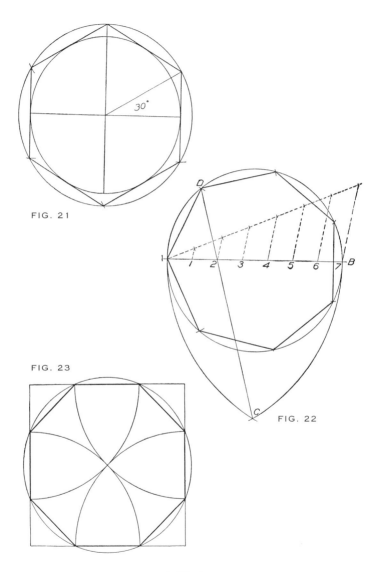

FIG. 21

30°

D

1 2 3 4 5 6 7 B

C FIG. 22

FIG. 23

EXERCISE 22

A hexagon circumscribed about a given circle (Fig. 21).

EXERCISE 23

A heptagon within a circle.

Construction: Draw a line making any angle with AB. Divide AB into as many equal divisions as the polygon has sides. With A and B as centers and AB as a radius, describe arcs at C. A line drawn through C and 2, cutting the circle at D, determines the length of one side of the heptagon. This method can be applied to any polygon. Use the formula in Exercise 20 to determine the size of the interior angle (Fig. 22).

EXERCISE 24

An octagon within a circle.

Construction: Within a circle of an assumed diameter divided into quadrants, draw bisectors. The circumference is now divided into eight equal divisions. Determine and locate the size of the interior angle using the formula given in Exercise 20 (Fig. 23).

EXERCISE 25

An octagon within a square (Fig. 23).

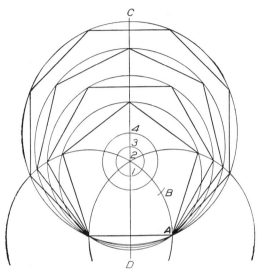

FIG. 24

EXERCISE 26

A combination of polygons on a given base of 1 inch.
Construction: Proceed as if marking out a hexagon. Bisect
the arc A-2. Trisect 2-B. With center 2, draw arcs cutting
through C and D at 1 and 3. Points 1 and 3 are centers of
circles circumscribing polygons of the required number
of sides (Fig. 24).

EXERCISE 27

Inscribe circles about the triangles given in Exercises 10,
12, 13, and 15.

EXERCISE 28

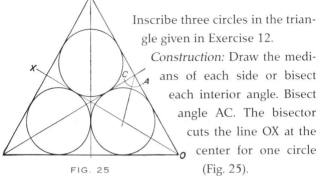

Inscribe three circles in the triangle given in Exercise 12.

Construction: Draw the medians of each side or bisect each interior angle. Bisect angle AC. The bisector cuts the line OX at the center for one circle (Fig. 25).

FIG. 25

EXERCISE 29

Five circles tangent to a given circle and each other inside or outside the given circle.

Construction: Divide the given circle into five equal parts and bisect each sector. The centers of each circle will be located on the bisector. Draw a tangent at the terminal of a bisector and extend it until it cuts a radial line. Bisect the angle this tangent makes with the radial and extend this bisector until it cuts AB at C, which is the center of one circle (Fig. 26).

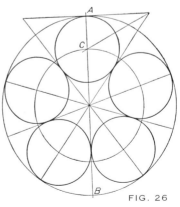

FIG. 26

EXERCISE 30

A circle tangent to a given circle and a given line.

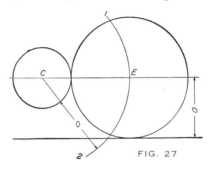

FIG. 27

Construction: With the radius of the required circle added to the radius of the given circle and C as a center, strike an arc 1-2. Draw a line parallel to the given line with the distance equal to the radius O of the required circle. This line and arc 1-2 intersect at the center E for the required circle (Fig. 27).

EXERCISE 31

A shaft $1\frac{1}{2}$ inches in diameter rotates within a ball bearing consisting of ten tempered steel balls. Make a drawing illustrating the size of the balls required (Fig. 28). Approximate.

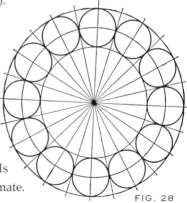

FIG. 28

Construction: Proceed as in Exercise 29 except make the tangent circles external to the given circle.

EXERCISE 32

The four largest circles that can be drawn within a square.

EXERCISE 33

A Maltese cross.
Construction: Draw two equal circles upon the two diameters of a given large circle and proceed as indicated in the drawing (Fig. 29).

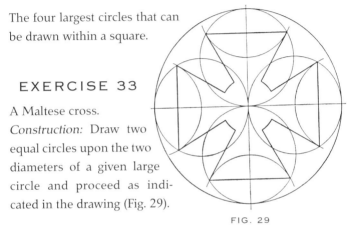

FIG. 29

EXERCISE 34

Draw a geometric border using the circle as a unit. Graphic design has historically been geometric in character (Fig. 30).

FIG. 30

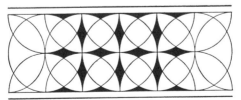

EXERCISE 35

Draw a box within an interlocking set of knots (Fig. 31).

FIG. 31

EXERCISE 36

Geometric monogram within a trefoil (Fig. 32).

FIG. 32

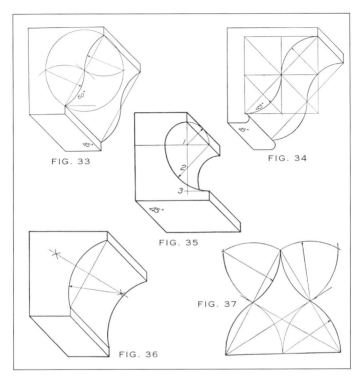

FIG. 33

FIG. 34

FIG. 35

FIG. 36

FIG. 37

EXERCISE 37

Moldings.

    1. Cyma Recta (Fig. 33)

    2. Roman Ogee (Fig. 34)

    3. Scotia (Fig. 35)

    4. Echinus (Fig. 36)

    5. Ogee Arch (Fig. 37)

EXERCISE 38

According to astronomers, the plane of the earth's orbit is elliptic. Draw an ellipse by two methods. Construct the upper half as follows : AC = $4\frac{1}{2}$ inches, DE = $3\frac{1}{2}$ inches, and DM = AB. M and F are centers of all the arcs on the ellipse. From C as the center, mark off on BC any number of points, 1, 2, 3, 4, 5, etc. With C-1 as a radius and M and F as centers, describe the arcs.

With A-1 as a radius and MF as centers, describe arcs intersecting C-1. These are points of the ellipse. Proceed until enough points are determined to locate the curve.

Construct the lower half by the circle method, the steps for which are self-evident from the illustration (Fig. 38).

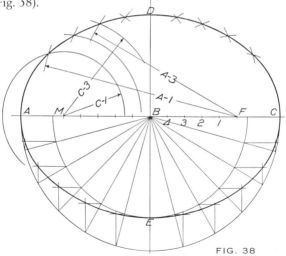

FIG. 38

EXERCISE 39

The trammel method.

Construction: On a small strip of cardboard, mark off the semi-minor and semi-major axes
equal to the dimensions of
Exercise 38. Move point
C so that it is always on
line AB and point E on
line DF. By changing the
position of the trammel
frequently, sufficient points
can be located at G, on the tram-

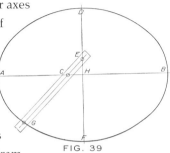

FIG. 39

mel, to determine a symmetric ellipse. Make GC = DH
and GE = AH (Fig. 39).

EXERCISE 40

Make a full-size drawing of the elliptic cam (Fig. 40).

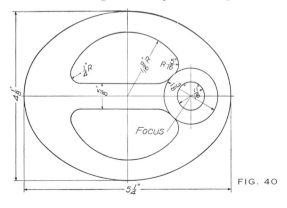

FIG. 40

EXERCISE 41

A point on a connecting rod of a stationary engine describes an elliptic curve in one revolution of the crank wheel. With B as the given point, mark out the desired curve. The construction for the mechanism may be omitted (Fig. 41).

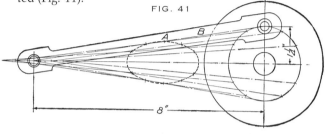

FIG. 41

EXERCISE 42

A five-point elliptic arch with three radii.

Construction: AB, the altitude, and CD, the span, are given. Mark off the major and semi-minor axes. With A as a center and AB as a radius, draw an arc through BE. Bisect CE at F and describe an arc with CF as its radius. CG = AB and is perpendicular to CD. G-3 is perpendicular to BC, and where G-3 intersects CD at 1 is a point of the first center of the ellipse. Where it cuts AB at 3 is another. Make AH = BK. With 3 as a center and 3-H as a radius, describe an arc through H. With C as a center and AK as a radius, strike an arc at N. With 1 as a center and 1-N as a radius, strike an arc at 2, which is another point of a center for the ellipse. With the construction duplicated on the right of

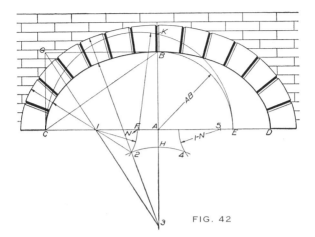

FIG. 42

the illustration, the remaining centers are determined. Points 1, 2, 3, 4, and 5 are the required centers, and all arcs and facing stones radiate from their respective centers (Fig. 42).

EXERCISE 43

A cycloid is a curve generated by the motion of a point on the circumference of a circle as it rolls along a straight line. The figure clearly illustrates the construction. Imagine the rolling circle to be the end of a cylinder (Fig. 43).

FIG. 43

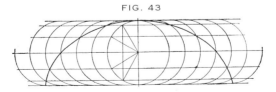

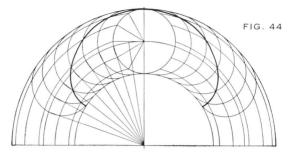

FIG. 44

EXERCISE 44

An epicycloidal curve is generated by the motion of a point on the circumference of a circle that rolls on the outside of a fixed circle (Fig. 44).

EXERCISE 45

A hypocycloidal curve is generated by the motion of a point on the circumference of a circle rolling upon the concave side of a circle (Fig. 45). If the diameter of the generating circle equaled the radius of the larger circle, the hypocycloid would become a straight line.

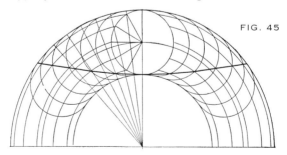

FIG. 45

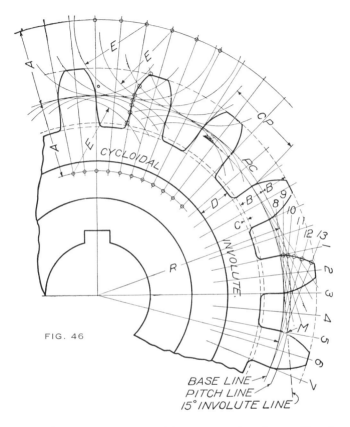

FIG. 46

BASE LINE
PITCH LINE
15° INVOLUTE LINE

These curves are used in constructing the profile of gear teeth. Fig. 46 is a draftsman's method of laying out the forms of teeth theoretically, the method toward the bottom being involute and that toward the top, cycloidal. Fig. 46a is a perspective sketch of the same forms from a

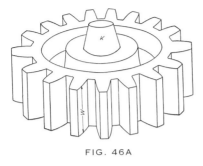

FIG. 46A

pattern made by a patternmaker in the shop. The size of the rolling circle, 2E, in determining the *epi-* and *hypo*cycloidal curves is not a fixed diameter; however, it is best to make it one-half the diameter of the pitch circle of the smaller of two engaging gears. In a problem where the diameter of PC, or 2 x r, and the number of teeth *n* are given, the circular pitch, which is the distance from one tooth to a corresponding point of another, CP, must be marked off first on PC.

The involute method is as follows: at the radial line 2, draw a tangent 8 where it intersects the base circle at 2. On this tangent, mark off the chord of the arc PC between radials 1 and 2. This is a point of the tooth's curve. At radial 3, repeat the above process, but mark off two chords of the arc on tangent 9 instead of one; on 10, three chords, and so on, until enough points are secured to define the desired involute tooth curve. Reverse the operations for the other side. CP/2 equals the width of the tooth or space for all purposes in drafting. The lower half of the tooth's profile is a radial line. The base circle is drawn tangent to the involute line of 15° through M.

The same principle is involved in laying out the cycloidal tooth; however, the chords of the arcs on PC are marked off on the arcs of the rolling circle. Above the line PC, the rolling circle generates the epicycloidal profile, or addendum, and below the line, the hypocycloidal or dedendum. A = E.

The following additional data are given for those who would like to specialize in gear teeth:

Addendum = depth of tooth above PC = 35 CP. $\Big\}$B
Dedendum = depth of tooth below PC = .35 CP.

Clearance at root of space = .05 to .1 of CP. $\Big\}$C

Actual thickness of tooth on PC = .45 of CP. $\Big\}$*
Actual width of space on PC = .55 of CP.

Backlash, or play between engaging teeth = .1 of CP.

Circular pitch (CP) = a tooth and space on PC; more commonly used than diametral pitch.

Diametral pitch (DP) = a certain number of teeth per inch of diameter of PC.

If DP = 1   CP = 3.1416.
      = $1\frac{1}{2}$ = 2.094.
      = 2   = 1.571.
      = $2\frac{1}{4}$ = 1.396.
      = $2\frac{1}{2}$ = 1.257.

A π (pi) relation exists between circular and diametral pitch: if π is divided by DP, the result will be CP; if π is divided by CP, the result will be DP.

*About equal in machine-cut gears

Let n = number of teeth.

CP = $(\pi D)/n$ when D = diameter of PC.

$n = (\pi D)/CP$

The thickness of rim D = .12 + .4 CP.

The width of face W, Fig. 46a, averages 2 to $2\frac{1}{2}$ CP.

The diameter of the hub = twice the diameter of the shaft.

The thickness of the web connecting the hub and the rim varies.

Arms are used on larger gears. Holes are often drilled through the web to lighten the weight without destroying the efficiency of the gear wheel. The length of the hub may be flush with the rim, but is usually $\frac{1}{4}$" or more longer.

The face of a tooth is the distance B above PC. The flank of a tooth is the distance B below PC. In Fig. 46a, K shows the position of the core print used in molding the hole for the shaft.

PROBLEM 1

Draw the front and side views of a gear wheel having 24 teeth and $2\frac{1}{4}$ DP, with an epicycloidal profile of teeth. Scale: full size.

PROBLEM 2

A pinion for a certain gear has 27 teeth. The CP is 1.571 inches. Draw forms of teeth by the involute method. Scale: half size.

Note: A pinion is the smaller of two gears acting together and should not have less than 12 teeth.

PROBLEM 3

Make a scale shop drawing of a pair of meshing gears of 8 DP. The driver should be a plain gear with 32 teeth, a 1" face, a 1" bore, and one hub $\frac{7}{8}$" long. The following gear should be a web gear traveling at two-thirds as many revolutions per minute (RPMs) as the driver. The follower should have a $\frac{5}{16}$" web, a $\frac{1}{4}$" rim or backing, and 2" hubs, one flush and the other $\frac{7}{8}$" long. Each gear should be held on the shaft by two kinds of fastenings. All unspecified dimensions and details are at the discretion of the draftsman, whose goal should be to make a mechanism of ordinary and reasonable proportions. The driver should be finished all over (FAO); the follower should be finished (f.) at the rim, the ends, and the outside of its hubs. Scale: full or double size.

Note: The profile of the teeth is not necessary for cut gears.

EXERCISE 46

An Archimedean spiral of one whorl.
Construction: With a radius equal to the rise of the spiral AB, and A as a center, describe a circle. Divide AB into as many equal divisions as the circle has been divided into sectors. Mark off successive arcs on the radials and draw in the curve. If a spiral of two whorls is desired, divide AB

into twice as many parts as for one whorl. This problem represents a cross section of the nautilus, a sea shell described by the Oliver Wendell Holmes poem "The Chambered Nautilus" (Fig. 47).

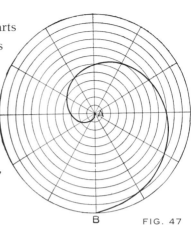

B

FIG. 47

EXERCISE 47

A heart plate cam. The construction for this common object may be derived from Exercise 46 and the figure. The bobbin winder on a sewing machine is one of several applications (Fig. 48).

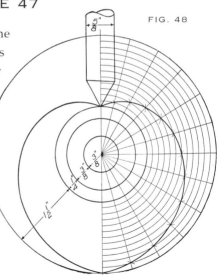

FIG. 48

EXERCISE 48

An involute spiral. This curve is developed by unwinding a string wrapped about a cylinder, the end describing the involute.

Construction: Mark off tangents at regular intervals to the cylinder. On the first tangent line, step off the chord of one arc. On the second tangent, two chords; on the third, three, etc. Draw the curve through the points (Fig. 49).

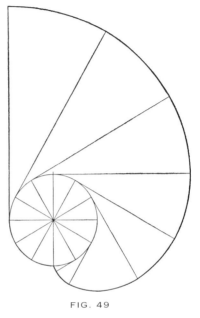

FIG. 49

The involute is used in defining the tooth curve of a gear wheel.

EXERCISE 49

A helix (Fig. 50). A helix is the combined vertical and horizontal motion of a point about a right line as an axis, with no two points of the curve lying in the same plane. The upper part of Fig. 50 shows this part laid out apart from its application to the screw thread.

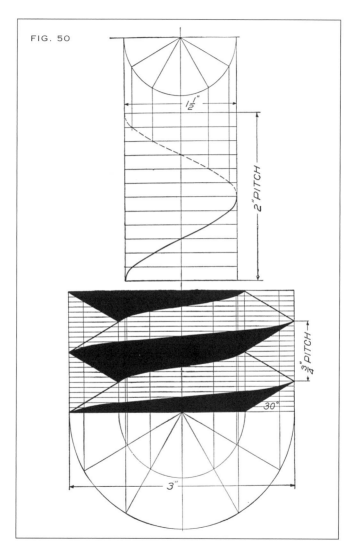

FIG. 50

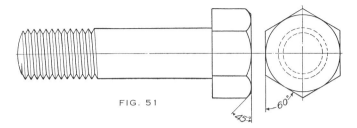

FIG. 51

Construction: Mark out the plan and elevation of the thread desired. Divide the half section of the plan into any number of equal parts and divide the pitch into the same number. The curves are obvious from the illustration, which is a single thread—the winding of one screw thread around the bolt cylinder. A double requires two threads parallel to each other; a triple, three, and a quadruple, four. Fig. 51 represents a conventional method of showing the single thread in practice. No attention is paid to the theoretical helical curve in drafting; however, it is essential to have a proper understanding of it.

EXERCISE 50

The Ionic volute. Fig. 52 is an illustration of the volute spiral of an Ionic capital from classic architecture (Fig. 55). In marking out such a curve, the methods in either Fig. 53 or Fig. 54 may be used. When AB is given (Fig. 52), make the eye of the volute one-sixteenth of AB and locate its center on the ninth division of AB. Divide the semidiagonal of the square into three equal parts and

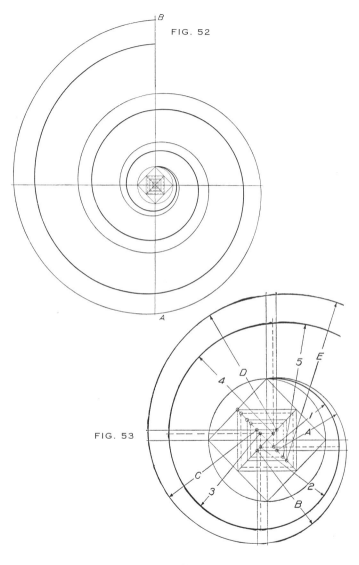

FIG. 52

FIG. 53

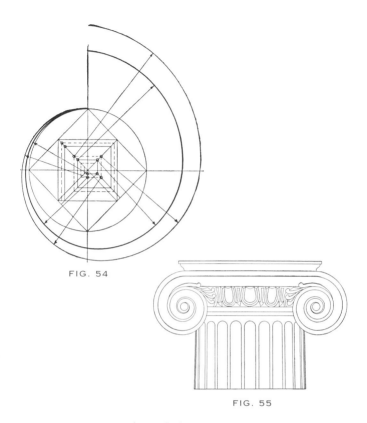

FIG. 54

FIG. 55

construct squares through these points, as in Fig. 53. Each corner of these squares is a center for quadrants of the outer spiral, starting with radius A. The inner dotted squares are drawn to pair within the first squares, a distance of one-third the space between the first series of squares. Proceed with the construction of the spiral following consecutive radii.

The second method is practically the same. The radii of both the inner and outer quadrants are taken on line CD. Follow the unbroken lines until the spiral is completed. The construction of the diagonal in beginning this method is the same as in Fig. 53. The offset diagonal is equal to one of the smaller spaces on the diagonal (Figs. 52, 53, 54, and 55).

EXERCISE 51

Draw a 1-inch bolt 3 inches in length. There are eight threads per inch. The angle of the Vs in the U.S. standard or sellers thread is 60°. Note the difference between a single- and double-V thread in the conventional layout. A square has half as many threads as a V of the same diameter. Show the length of the bolt from the underside of the head's edge to the end of the cylinder (Figs. 51 and 88, page 84).

Ornamental and decorative art implies the use of geometry in laying out designs and patterns in stained and art glass, carpets, wallpaper, oilcloth, borders, ornamental iron and woodwork, china painting and pottery, as well as carving, carpentry, cabinetmaking, floor tiling, and bookbinding. Franz Sales Meyer, in his *Handbook of Ornament,* said: "In medieval times these geometric constructions developed into practical artistic forms as we now see them in Moorish paneled ceilings and Gothic tracery." One may also see traces of geometric design in the tattoos and decorations on the implements of American Indians.

Geometric motifs may be obtained from the flowers. The trillium and daisy are illustrations of the triangle and circle; the columbine and lilac, of polygons. These may be arranged into rosettes, borders, and stencils using the circle as a unit. The illustration of the trefoil, Fig. 32 (page 40), is a design of a monogram of an appropriate initial. Problems pertaining to decorative design are beyond the scope of this course, but are reserved instead for a later work.

CHAPTER IV

WORKING DRAWINGS

Reference has previously been made to the value and importance of having working drawings in the shops. No reliable workman should attempt a new problem without first making use of a working drawing. A plan is the appearance of the top of an object when observed from above. An elevation is the appearance of the object when observed from the front or side. No first-class foreman should permit a workman to begin a task without having ample directions in plan and elevation. Any ordinary problem can be clarified in the shops with three views of the object. Occasionally, a very irregular object requires special views, but for the purposes of this course, these will be omitted.

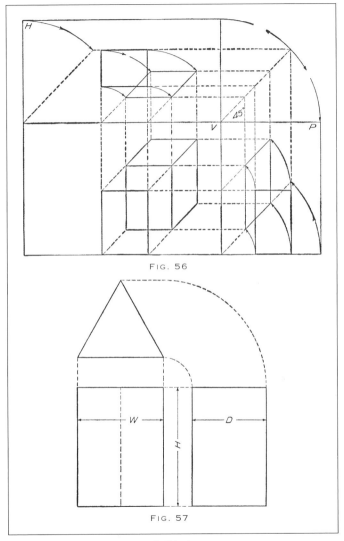

FIG. 56

FIG. 57

Place a model within a glass case and trace a plan view on its upper surface. Likewise, trace a view of its front and side. Now open each plane until top, front, and side lie in one flat surface, as in Fig. 56. This is the working drawing, or three projected views. Both planes H and P revolve through 90°. Note the references to height (H), width (W), and depth (D), together with the method of obtaining them from one view and carrying them to another, in Fig. 57.

If a glass case with hinged sides is not conveniently acquired, select and invert a good cardboard shoebox over any geometric model. Sever all but the front edges, which will serve as hinges. Outline on the surfaces the shape of the several views, and then cut out the outline from planes H, V, and P. A tin can, cardboard box, prism, or any other simply shaped object will serve well as a drawing exercise. Many problems should be drawn so the fundamental principles of projection that underlie the working drawing can be studied and understood. While geometry exercises form the basis of all mechanical drawing, the details and principles of the workman's drawing are used most often and are the practical application of constructive drawing.

The craftsman, patternmaker, machinist, and carpenter must each have a definite plan or idea prescribed before him in the form of a blueprint working drawing. The first thought of any constructive character must

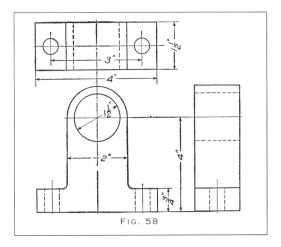

FIG. 58

always first appear on paper, and the common means of
that representation is the kind of drawing described
above (Fig. 58).

EXERCISES

1. A rectangular prism. Draw three views (Fig. 59).

2. A pentagonal plinth. Draw three views (Fig. 60).

3. A bushing pattern. Draw two views (Fig. 61).

4. An angle iron. Draw three views and dimensions
   (Fig. 62).

5. A cast-iron block. Draw two views and dimensions
   (Fig. 63).

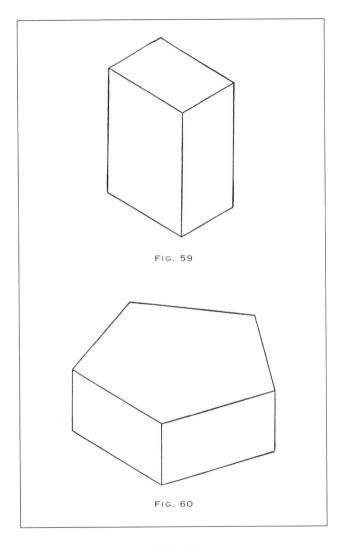

FIG. 59

FIG. 60

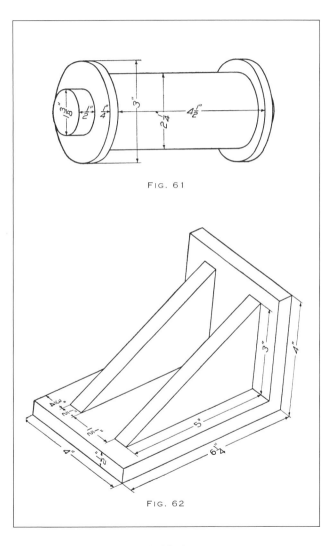

FIG. 61

FIG. 62

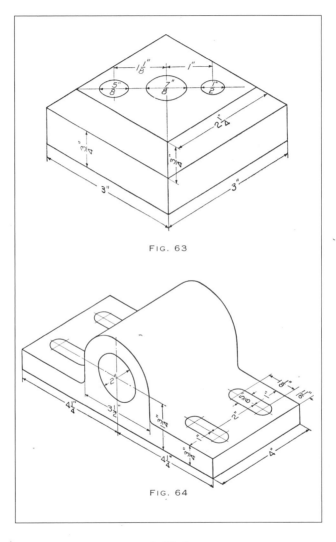

FIG. 63

FIG. 64

6. A pillow-block bearing. Draw three views and dimensions (Fig. 64).

7. A tool post holder. Draw three views and dimensions (Fig. 65). Scale: half size. Opening of slot: $4\frac{1}{2}$" long.

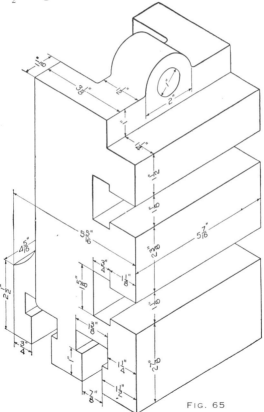

FIG. 65

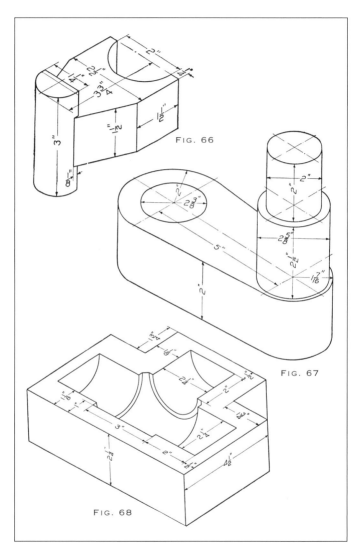

FIG. 66

FIG. 67

FIG. 68

8. A rocker. Draw three views and dimensions (Fig. 66).

9. A crank arm. Draw two views and dimensions (Fig. 67).

10. A core box for pipe tee. Draw three views and dimensions (Fig. 68).

11. A coupling. Draw two views and dimensions (Fig. 69).

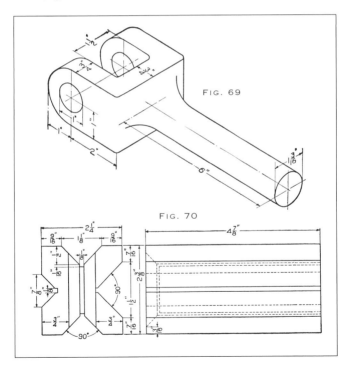

FIG. 69

FIG. 70

12. A V block. Draw three views and dimensions (Fig. 70).

13. A pattern for a shaft bearing without the cap. Draw three views (Fig. 71).

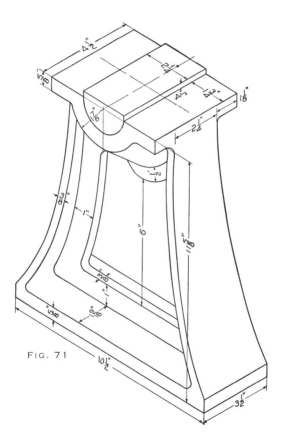

FIG. 71

[CHAPTER V]

CONVENTIONS USED IN DRAFTING

Conventions, as explained in Chapter I, are customary methods or symbols established by usage and precedent and are generally employed for the sake of uniformity and convenience the world over. Their use and convenience will be readily understood by the student.

a. Circles require two centerlines and must always be shown.

b. Invisible edges are shown by a series of $\frac{1}{8}$" dashes with $\frac{1}{16}$" white space between them.

c. Visible edge lines take precedence over invisible lines when they coincide.

d. Dimension lines are continuous, very light, and broken only for dimensions, near the center of the dimension line.

e. Dimensions should read at right angles to the dimension lines in the working drawing.

f. Sharp, precise arrowheads should attach to the ends of each dimension line.

g. The summation or aggregate of several dimensions tending in any one direction should be shown separately so workmen do not err in calculating overall sizes of stock required for the finished product.

h. Projection lines are light single dashes of any desirable length extending from view to view that facilitate the placement of the dimensions. They should not touch the projections of the figure.

i. When the space to be dimensioned is too crowded, use arrows outside that space directed toward the figures to be dimensioned.

j. Draftsman's figures are to be used for all numerals. Fractions must be must be common shop units, such as $\frac{1}{32}$, $\frac{3}{16}$, $\frac{5}{8}$, $\frac{1}{2}$, etc., and never $\frac{1}{3}$, $\frac{1}{6}$, $\frac{1}{10}$, or $\frac{1}{12}$. The bar separating the numerator from the denominator should always be drawn horizontally to avoid any possible mistake in reading a dimension.

k. Section planes use the same convention as centerlines.

l. All material edges cut by a plane are represented by solid lines.

m. Crosshatch adjacent pieces in an assembly of parts at right angles or in different directions. Do not space the hatch lines too closely.

n. Dimensions placed to the right and between views are less likely to be overlooked by workmen.

o. Do not permit dimension lines to cross each other.

p. Show dimensions between centerlines and finished surfaces. They are very important.

q. Sections are shown to clarify the hidden details of construction. They should occur frequently and be properly located in complex drawings.

r. Do not repeat dimensions except in a very complicated drawing.

s. Always place full-size dimensions on the drawing, no matter what scale is used.

t. Locate the front elevation first.

u. Invisible parts behind sections are never shown.

v. Bolts, shafts, and screws are never sectioned. A broken cross section of a bolt or shaft should show the convention of the material.

w. Show diameters in preference to radii.

x. Never crosshatch over dimensions.

y. Arcs of circles and curves should be drawn before straight lines that adjoin them.

LETTERING

One of the most important features of any drawing, and one of the most neglected by students and amateur draftsmen, is the neat appearance of every detail. These details consist chiefly of letters, figures, notes, titles, scales, stock lists, and lists of materials—data that, if executed neatly and with precision, improve the appearance of what might otherwise be a poor drawing one hundredfold.

Although they may be expert in draftsmanship, architects and mechanical engineers are nonetheless obliged to letter well in order to retain their positions in many companies. This is why technical schools emphasize the quality of their students' work. Notebooks, examination papers, programs, and blueprints receive better consideration when their titles are well lettered than when scribbled in some unreadable characters.

Fig. 72 is an exercise containing all the letters of the alphabet; such a sentence is usually part of the exercises in a typing class. Lowercase (or small) letters are shown in Fig 73.

Pencil all letters freehand for guides with a 2H pencil, and submit for approval. Ink with a ruling pen.

THE FOLLOWING IS A GOOD EXERCISE
AND CONTAINS ALL THE LETTERS OF
THE ALPHABET:—
"THE QUICK BROWN FOX JUMPS OVER
THE LAZY DOG."

HARD PRACTICE IS A GOOD MASTER.
THE DRAFTSMAN'S FIGURES ARE AL-
WAYS USED. 1 2 3 4 5 6 7 8 9 0 $\frac{1}{2}$" $\frac{13}{16}$"

ALL LETTERS AND FIGURES SLOPE
HALF THEIR HEIGHT, OR ABOUT 30°.

FIG. 72

This lower case style is a very
popular form of letters for notes,
titles, stock-lists, bills of material, etc.

"A quick brown fox jumps over
the lazy dog."

Stem letters are $\frac{3}{16}$ths of an inch
high. Use a ball-pointed pen #506 or
#516.

FIG. 73

(*Note:* Do not mistake a No. 2 pencil for a 2H pencil. Any good stationer can explain the grading of pencils. If graph paper is not available, rule a sheet into $\frac{1}{8}$" squares and draw slope lines about 30° from a vertical, as in Fig. 73.)

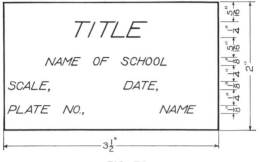

FIG. 74

All drawings should be titled properly (Fig. 74). Note that the most important part of the title is the object of the drawing, which is therefore more prominent than the rest. A title should be balanced so that one side does not seesaw or appear heavier than the other. The titles need not be circumscribed by a 2 x $3\frac{1}{2}$" boundary as this one is, but the proportion of the spacing between lines and heights should follow this illustration's. Place the title plate in the lower right corner, about $\frac{1}{2}$" from the margin line.

Draw all downward strokes first, then curves or intervening lines as indicated in Fig. 73. Oval letters and figures are constructed on the form of the letter O. Make the letter O and then modify it to suit the shape of the desired letter or figure.

Practice lettering the exercise given in Fig. 75 until proficiency is assured.

THIS FREEHAND GOTHIC STYLE OF LETTERS SHOULD BE USED ON ALL DRAWINGS NOT ARCHITECTURAL OR TOPOGRAPHICAL.

MAKE ALL LETTERS UNIFORMLY HIGH, $\frac{1}{8}$TH INCH IF LOWER CASE AND $\frac{1}{4}$TH IF CAPS. THIS STYLE IS $\frac{1}{8}$TH CAPS.

DRAW LIGHT GUIDE LINES FOR THE SLOPE AND HEIGHT. WIDE LET-TERS ARE BEST. SPACE BETWEEN WORDS SHOULD NOT BE LESS THAN $\frac{1}{4}$TH INCH, NOR MORE THAN $\frac{3}{8}$THS. KEEP LETTERS IN EACH WORD COMPACT.

GOOD LETTERING ENHANCES THE APPEARANCE OF ANY DRAWING.

NEATNESS AND LEGIBILITY ARE VALUABLE ASSETS IN MECHANICAL DRAWING.

DO NOT USE A VERTICAL STYLE.

FIG. 75

A final style of lettering is architectural, shown here in Fig. 76.

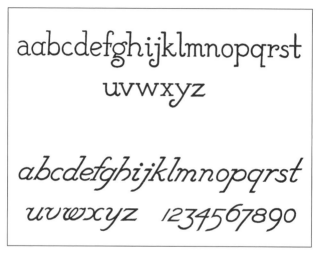

FIG. 76

CHAPTER VI

MODIFIED POSITIONS OF THE OBJECT

Suppose the prism in Fig. 59 (page 64) were revolved about a vertical axis; that is, an axis perpendicular to the H plane. Looking down on the prism in Fig. 59, how would the plan be drawn if the object were revolved 30° about a vertical axis? Do the altitude and the construction of the plan view alter in such a revolution? Use a cardboard box or other simply shaped object as a model until each step in the thinking process is clear. Now draw the remaining views. Through how many degrees does the earth revolve? Can anything revolve? Any point of the object always moves in a plane perpendicular to the axis. There are 360 degrees in a circle.

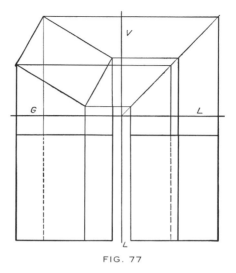

FIG. 77

PRINCIPLE 1

When an object revolves about a vertical axis (perpendicular to H), its plan view is not altered in shape, but only in position, and its height remains the same (Fig. 77).

Note: No distinction is made here between vertical and perpendicular, as the H plane is always considered horizontal.

Find the projections of the plinth in Fig. 60 (page 64) when it is revolved through an angle of 30° about a side axis (i.e., parallel to the profile or side plane).

Each point of the object revolves in a circular plane, or path, through 30°, about the side axis, which can only be seen as a point from the front. Therefore, the front elevation will not be altered in construction from its original and natural position, but its position will be 30° inclined to the base upon which it originally lay.

Find its remaining projections. Looking down on the plinth in Fig. 60, as on the prism, when it is revolved about a side axis (perpendicular to V) the depth or thickness does not alter, but the construction of the plan changes.

PRINCIPLE 2

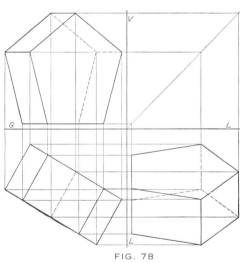

FIG. 78

When the object revolves about a side axis (perpendicular to V) to the right or left, its front elevation does not change (except for its position), and the depth (D) remains the same (Fig. 78).

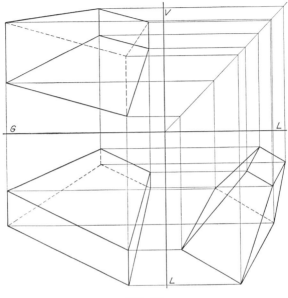

FIG. 79

Find the three views of the frustum of the pyramid
(Fig. 79) when it is revolved about a front axis through an
angle of 30° forward or backward (perpendicular to P).
First observe the position of the object from the profile
or right-side plane. Then tilt the side elevation forward or
backward to make the revolution about the front axis,
which may be seen as a line parallel to the ground line
(GL) and passing through the center of the model. The
lower edge will make the required angle with the base
upon which the object stands.

Look down on the prism in Fig. 59 (page 64) once again. The width of the object does not alter when it is revolved forward or backward. Draw the three projections when it is so revolved; begin with the side, then draw the front, and finally, the plan. In this case all points of the object revolve in planes, which the observer can only see as straight lines parallel to the vertical line (VL).

PRINCIPLE 3

When the object revolves about a front axis (perpendicular to P) forward or backward, its side elevation does not change except in its position, and the width remains the same.

CHAPTER VII

THE DETAILED WORKING DRAWING

A machine is a composition of many parts. Each part performs a certain function and bears a close relation to adjacent parts. If a mechanic desires to make a machine, he must organize the parts perfectly so that the required work creates a minimum amount of friction. When each part is made in the shops, each specific detail is worked out separately. For example, a connecting rod carries power directly from the cylinder through the piston to the drivers of a locomotive. This object is one detail of a stationary engine. A detailed working drawing makes it as easy as possible for the mechanic to imagine the size, shape and, to some degree, the relative position of the parts, and to construct each without confusion.

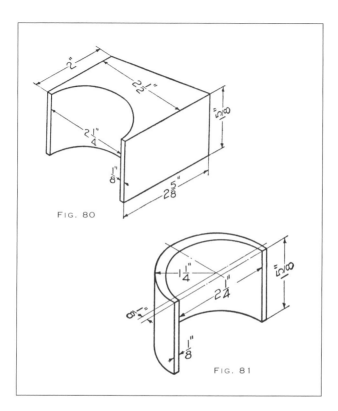

FIG. 80

FIG. 81

From the illustrations above, make a working drawing of each separate part and then fit them together in an assembly drawing. Figs. 80 and 81 are parts of a bearing that surround the cross-head pin and fit in the left end of the rod (Fig. 87). Fig. 82 is a tapered key block used to take up wear and is located behind Fig. 80. Figs. 83 and 84

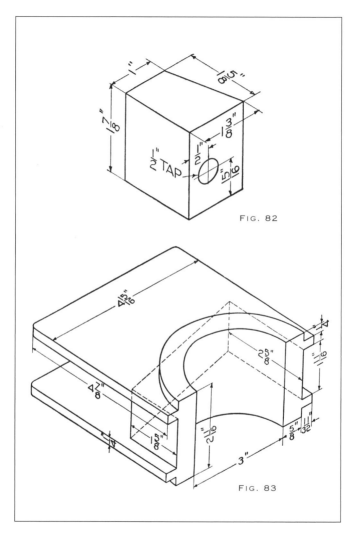

FIG. 82

FIG. 83

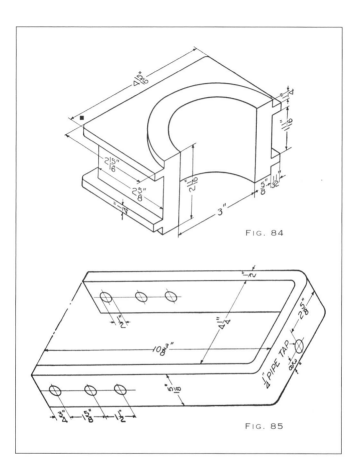

FIG. 84

FIG. 85

are parts of the bearing that fit around the crankpin and are located on the right end of the rod (Fig. 87). A strap (Fig. 85) holds these two parts together with another

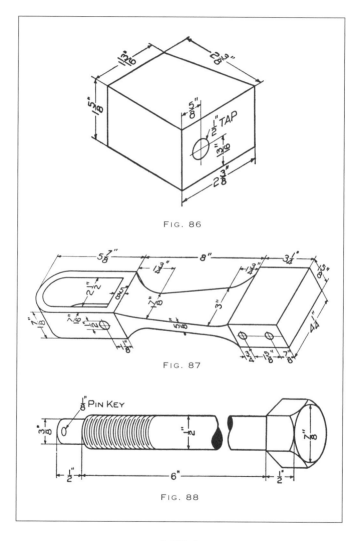

FIG. 86

FIG. 87

FIG. 88

STOCK LIST

MARK	NO. WANT	NAME	MATERIAL	REMARKS
80-81	1	BEARING	PH.BRONZE	2 PARTS
82	1	KEY-BLOCK	STEEL	F.A.O.
86	1	" "	"	"
83-84	1	BEARING	PH.BRONZE	FINISH BABBITT
85	1	STRAP	STEEL	F.A.O.
87	1	ROD	"	"
88	2	BOLTS	W.I.	$6"x\frac{1}{2}"D$
	1	"	"	$3"x\frac{1}{2}"D$
	2	"	"	$1\frac{3}{4}"X\frac{1}{2}"D$
	1	"	"	$1\frac{1}{2}"X\frac{1}{2}"D$

FIG. 89

tapered key block (Fig. 86) by means of $\frac{1}{2}$" bolts. Two $\frac{1}{2}$" bolts, each 6" long (Fig. 88) fasten the strap to the rod. Each tapered key block has two $\frac{1}{2}$" bolts, one on each side of the strap. This makes six bolts in all. The hole on the end of the strap is for oil. Copy the stock list in Gothic slant letters (Fig. 89).

When the drawing of a machine problem is completed, it is first sent to a patternmaker, who makes a model in wood from it. He must know from experience how much extra material to allow for shrinkage as the casting cools; how much to allow for polishing, or finishing; and how much to taper for draft when the pattern is

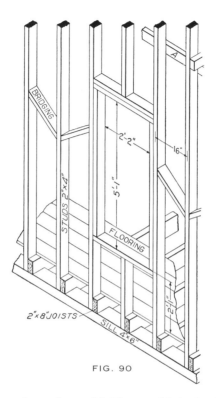

FIG. 90

withdrawn from the mold. The machinist is obliged to follow the specifications thereon, regardless of what he might think ought to be done. This places any responsibility for errors upon the draftsman.

A detailed working drawing is as necessary for the carpenter and cabinetmaker as for the machinist and engineer, and no one in the woodworking or building trades should attempt a task requiring skill and accuracy without it.

Clear, specific details must accompany the plans and elevations for the framework of a cottage. Studs, sills, rafters, sashes, joists, and so on should be located to provide the greatest service. Fig. 90 shows an isometric representation of a framing detail and, although not in accordance with the orthographic working drawing, provides the untrained eye with a better idea of the construction. Note the dimensions between members, size of stock, and joinery. The

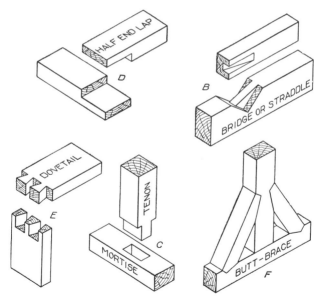

FIG. 91

plate for the second floor, at A, is usually set an inch into the studding. This is called a gained joint. Other forms of joints are miter, tongue-and-groove, open tenon with key, bridge (or straddle) (B), mortise-and-tenon (C), half-end lap (D), dovetail (E), and butt-brace (F) (Fig. 91). Each has a special purpose.

EXERCISE 1

Make an assembled drawing—plan and front elevation—of the framing details suggested in Fig. 90, and dimension properly. Scale $1\frac{1}{2}$" = 1' 0".

EXERCISE 2

Make a drawing of the stair detail in Exercise 1. Risers, 7"; tread, 10" wide; balusters, 2" square with space equal to the width of baluster (Fig. 92).

EXERCISE 3

Make a working drawing of the forms of joints used in joining represented in Fig. 91. Dimension it. Use stock sizes of materials. Scale: half size.

EXERCISE 4

Make a floor plan of your home, a barn, or a schoolroom, and show all appointments. Scale: $\frac{1}{8}$" = 1' 0". (Small details are usually drawn larger or to full scale.)

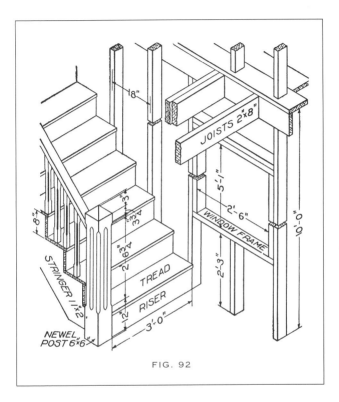

FIG. 92

EXERCISE 5

Make an architect's plan for the upper five-room apart-
ment in a modern three-story building. Use customary
architectural conventions to show all that is necessary and
usual. Scale: $\frac{1}{4}$" = 1' 0". Outside dimension: 22' 6" x 36'.

PATTERN-
WORKSHOP
DRAWINGS

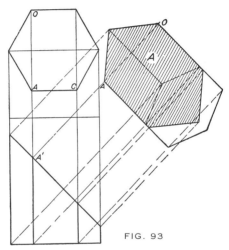

FIG. 93

One of the most useful applications of working drawing is to lay out patterns, or developments. The theory of such drawings is found in the study of descriptive geometry, which all architects and engineers are required to know something about and which is extremely useful to draftsmen, although its study is often avoided.

A thorough knowledge of the principles of pattern-making enables the tinsmith or sheet metal worker to lay out very complicated patterns in a very simple geometric manner, and consequently save time and material. Cutting a pattern as economically as possible requires foresight—something the typical patternmaker fails to exercise. Scraps may often be used to as good or better advantage than new sheets if conservatively and thoughtfully cut, and in all kinds of work stock should be ordered so that a minimum amount of waste is left.

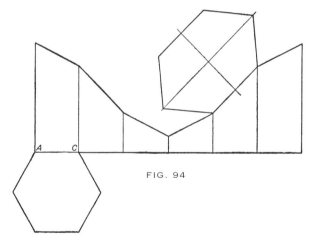

FIG. 94

A pattern is a plane surface representing the unfolded sides of an object equal to the perimeter of its right section, the width equal to the altitude of the object or, rather, the true length of its lateral surface.

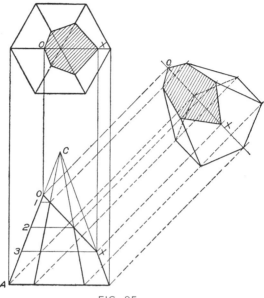

FIG. 95

Develop the surface of a cylinder or prism upon a sheet of bristol paper, allowing $\frac{3}{8}$" for lap. Glue the lap and fasten it together for a facsimile of the original. Add bottom and top.

In beginning a problem, it is usually only necessary to draw the plan, front elevation, and an auxiliary sectional view to show the true size of the cut section. For example, in Figs. 93, 94, and 95, the object is projected up into the plane of the paper to obtain the auxiliary view. After the pattern is drawn, it is transferred from the manila or bristol paper to the metal by pricking points with a sharp punch along the contour of the pattern, allowing for lap and seam. The double edge shown on the development of the quart measure is for the lock seam shown at A in Fig. 100.

PARALLEL METHOD

EXERCISE 1

Develop a truncated hexagonal prism (as shown in Figs. 93 and 94) using any suitable dimensions.

Construction: Draw the plan, elevation, and sectional view, as at A. The width of the section and base is the same as the depth of the prism transferred from the plan view. In the layout, the various heights of the linear elements of the prism are marked off on corresponding parallels in a straight line equal in length to the perimeter of the base.

EXERCISE 2

Develop a truncated hexagonal pyramid (as shown in Figs. 95 and 96) to suitable dimensions.

Construction: Obtain the projections and sectional view as in Problem 1. To obtain the development, first find the slant height of the pyramid. The exterior edges are parallel to the vertical plane; therefore, their true lengths must be seen at AC. With a compass set with AC as a radius, describe an arc. Mark off the perimeter of the base on

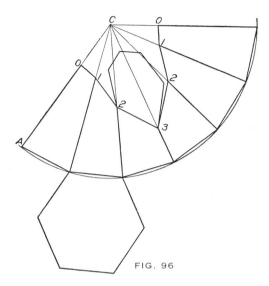

FIG. 96

this arc and join all points with radial lines to the center of the arc. Step off the true length of each cut element—0, 1, 2, and 3—shown projected on AC; then join as in Fig. 96. To complete the pattern, add the section and the base.

EXERCISE 3

Develop a frustum of a rectangular pyramid whose base is 2" x $1\frac{1}{4}$" and whose altitude is 3".

EXERCISE 4

An irregular cone is projected in Fig. 97. Develop by radial lines as in Fig. 96, but find the true length of each element separately.

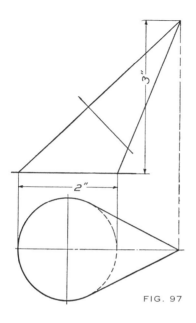

FIG. 97

EXERCISE 5

Given the front elevation of a $1\frac{1}{2}$" cylinder (Fig. 98), draw the plan and develop.

EXERCISE 6

Draw the pattern of a 1-quart measure whose upper base has a diameter of 3" and whose lower base has a diameter of 5" (Figs. 99 and 100). Find the altitude. (*Note:* This problem involves a principle of mensuration. Use either dry or liquid measure.)

$$\frac{V}{(\pi R^2)} = A \quad \text{or} \quad \frac{V}{D^2 (.7854)} = A$$

where

V = the volume or solid contents

and

A = the altitude.

This is approximate. To be exact, the formula should be stated as follows:

$$\{ a + b + \sqrt{ab} \}\frac{h}{3} = V$$

when

a = area of upper base.

b = area of lower base.

h = height or altitude.

There are 231 cubic inches in a liquid gallon and 2150.42 cubic inches in a bushel.

To find the altitude of a cone's frustum, substitute the known value of V, the volume, and solve for h as in any equation.

Fashion a suitable strip for a handle allowing a $\frac{3}{8}$" lap for edges. Fig. 99 shows the lap over a wire at the top of the cup. A customary rule for lap is four times the thickness of metal plus twice the diameter of the wire.

EXERCISE 7

An irregular triangular pyramid has an altitude of $4\frac{1}{2}$", sides of 2" or more in length, and all lateral edges

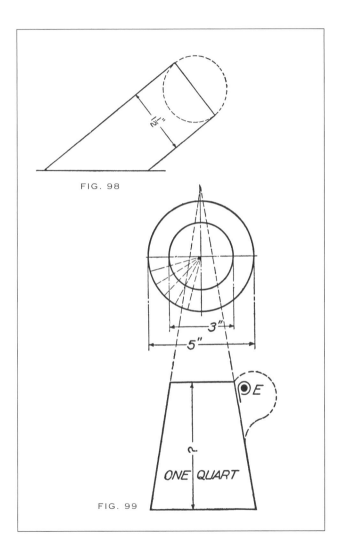

FIG. 98

3"

5"

E

ONE QUART

FIG. 99

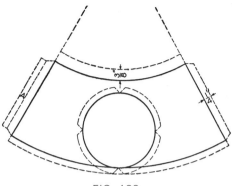

FIG. 100

oblique to all planes of projection. Two of its lateral edges and its base are cut by a sectional plane perpendicular to V and oblique to H. Draw the three orthographic views and the true size of the section and develop them (Figs. 101 and 102).

EXERCISE 8

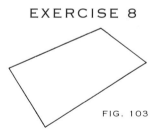

FIG. 103

An irregular oblique quadrilateral prism has a right section resembling Fig. 103. Its axis is inclined 30° to the right of its base, which is horizontal. A plane inclined 60° to the left of its base cuts all the lateral edges of the

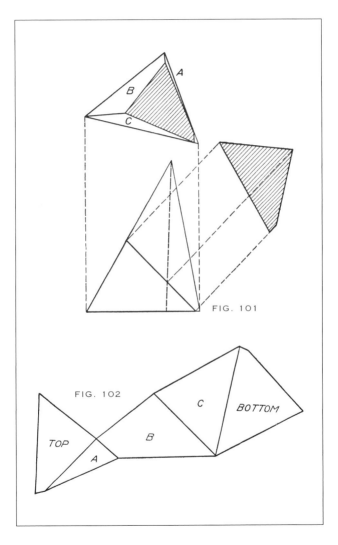

A

B

C

FIG. 101

FIG. 102

TOP

A

B

C

BOTTOM

prism. Draw the three projections and the auxiliary or sectional view. Develop, adding the base and sectional view. Use suitable dimensions.

EXERCISE 9

Draw three views of a regular vertical pentagonal pyramid its with apex above its base. The rear edge of the base is inclined 15° to the vertical plane of projection, V, the left end of this edge to be nearest V. The diameter of the circumscribing circle of the base is 2" and the altitude, 4". The pyramid is cut by a plane perpendicular to V at an angle of 60° to its base. Show the line of intersection in three views, make a sectional view, and develop either truncated part.

EXERCISE 10

A circular ventilator projects through a gambrel roof as shown in Fig. 104. Work out the line of its penetration with the roof planes. Develop the ventilator top and also the roof planes, showing the line of penetration. Scale 1" = 1' 0".

EXERCISE 11

Develop a truncated right cone based on the illustration on the facing page (Fig. 105).

EXERCISE 12

Develop a cylinder when R = $\frac{3}{4}$" and L = 3".

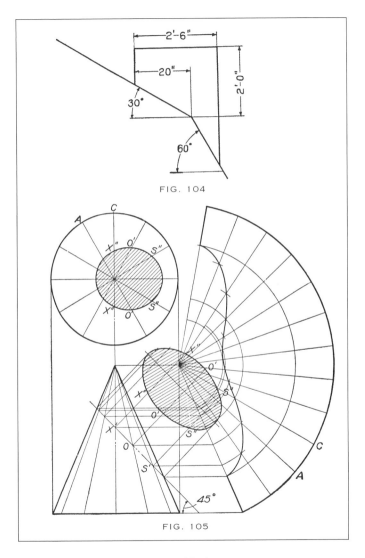

FIG. 104

FIG. 105

Any development of a geometric form is a mathematical process and hence should receive some such consideration.

The following formulas should be committed to memory:

$2\pi r$ = the circumference of a circle.

πr^2 = the area of a circle.

$(\pi r^2)L$ = the volume of a cylinder when
   L = altitude.

$(2\pi r)L$ = the lateral surface of cylinder.

$(\pi r^2)L/3$ = the volume of a cone.

$(2\pi r)S/2$ = the lateral surface of a cone when
   S = slant height.

$6(xy/2)$ = the area of a hexagon when x = one
   side of the polygon and y = the apothem.

Note: The apothem of a polygon is the perpendicular distance from the center of a polygon to one of its sides.

$6(xy/2)L$ = the volume of a hexagonal prism.

$6(xL)$ = the lateral surface of a hexagonal prism.

$(2\pi r)D$ = the lateral surface of a sphere when
   D = diameter.

EXERCISE 13

Develop a cone when R and the volume are given.

EXERCISE 14

The area of an octagon is 24 square inches. $X = \frac{3}{4}$".
Develop full size.

EXERCISE 15

$Y = \frac{1}{2}$", $L = 2\frac{1}{2}$". Develop.

Note: This problem involves the geometric construction of a hexagon without a circle before a development can be made. Using different data, originate and solve other problems.

EXERCISE 16

Fig. 106 shows the form of a sheet-metal hood for a forge. Scale: half size.

FIG. 106

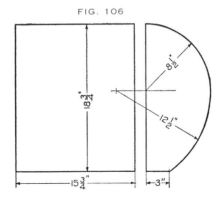

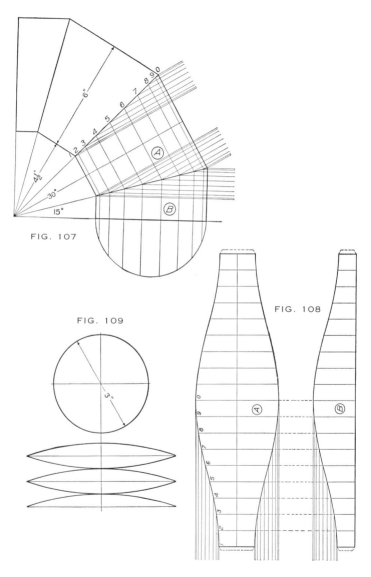

FIG. 107

FIG. 109

FIG. 108

EXERCISE 17

Fig. 107 gives the front elevation of a 6" stove pipe elbow, and Fig. 108 the development of a large and small section.

EXERCISE 18

Develop a 3" sphere using the "orange peel" method of turning a three-dimensional object into two-dimensional shapes (Fig. 109).

EXERCISE 19

Secure a good model of a funnel and draw the pattern.

METHOD OF TRIANGLES

EXERCISE 20

A tapering ventilator collar.

Many problems are impossible to develop by the orange peel method because their surfaces are warped. A warped surface cannot be laid out geometrically, but it can be constructed approximately by means of triangles. The moldboard of a plow, the "cowcatcher" on a locomotive, a marine ventilator funnel, a grain elevator spout, a cow's horn, a smokestack, a football, and similar objects with irregular surfaces can only be approximated by the triangle method.

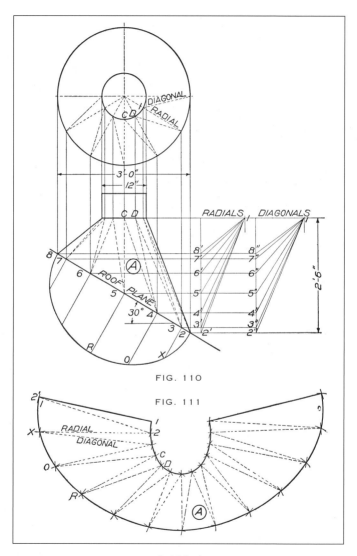

FIG. 110

FIG. 111

Construction (Fig. 110): Mark off small triangles on the projections of the figure at regular intervals, determine their true size, and lay them adjacent to each other. This will constitute, as closely as possible, a working pattern. The base of each triangle is shown in the plan view; the altitude, in the elevation.

The hypotenuse of any right triangle is easily determined when two of its sides are known.

Construction (Fig. 111): Mark off at any convenient place a radial and select the longest. At the lower end, strike an arc equal to X-2 on the sectional view of the roof plane. With a center at 2' (Fig. 111) and a radius equal to the length of the first diagonal 1-2", intersect the small arc 1-2". Small arcs are equal to C-D. With a center at 2" and radius 1-3, strike an arc at X, then mark off a second radial on either side of the first radial, 1-2'. Repeat until all the radials and diagonals have been used. Any warped surface may be developed in this manner.

EXERCISE 21

Select something from the kitchen that must be laid out in pattern and make a development to scale. Try a dust-pan, roasting pan, coffee urn, colander, or teapot. Use any method or combination of methods but be sure to determine whether the surface, or any part thereof, is warped (Fig. 97, page 99).

EXERCISE 22

Develop a truncated irregular cone by triangles.

Construction: This method is more mathematically exact than the previous methods, and it is often used to verify the radial line method. In geometry, a point revolves about an axis in a plane perpendicular to the axis. This holds true here, for the upper edges of the truncated section revolve in paths perpendicular to the lower edge of the base. The length of the radius of revolution is determined by constructing a right triangle one side of which always equals the distance AX from the horizontal projection of point A to the axis 1-2 in the plan, and the other side of which equals the perpendicular altitude A'-D' from the vertical projection to the plane of the base. The hypotenuse must equal RX. Connect 1-R, which is the true length of 1-A. This exercise should be repeated until it is clear. It is a graphical explanation of the same process in mensuration.

A second, closely related method, is to find the true length of each edge of the truncated pyramid and mark off these true lengths on the paths of revolution as drawn through the upper points A, E, F, G, etc., from points of the base X to R. The method of finding the true length of 1'-A' is shown by revolving 1-A parallel to GL and projecting to the base of the pyramid 1', then moving to its revolved position and also to A'. The true length of 1'-A' is now shown at 1"-A' (Fig. 113.)

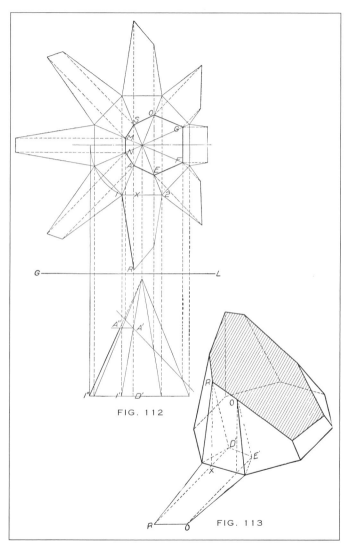

FIG. 112

FIG. 113

PENETRATIONS

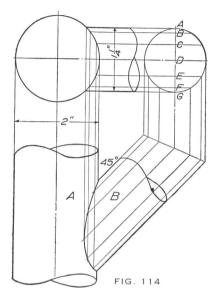

FIG. 114

When one object intersects or penetrates another, the line of intersection of the two is indented where they meet. To determine the pattern this line must always be geometrically located, as in Fig. 114, and herein frequently lies a difficult problem if the subject of working drawings and projections has not been thoroughly mastered.

DEVELOPMENT BY PARALLEL PLANES

EXERCISE 1

Fig. 114 is an illustration of two intersecting pipes. First, draw the plan and front elevation. Conceive a series of parallel planes, A, B, C, D, E, F, and G, cutting through both pipes and parallel to the front elevation. Each plane cuts two elements from each pipe, and all four elements lie in the same plane. In this case, two elements of pipe B penetrate one element of pipe A. Determine the projections of each element thus cut; where they intersect is a point of penetration.

To develop either A or B, lay out the perimeter of a right section with the height of the pattern equivalent to the length of the elements from the end of the cylinder to the line of penetration. A right sectional view shows the shortest possible circumference or perimeter of the object and is determined by a plane perpendicular to the axis of the figure.

EXERCISE 2

Fig. 115 represents a small rhombic prism penetrating a larger one, the top of each being a square in plan. Establish the lines of penetration in both plan and elevation, lay out the development of the smaller prism, and develop the hole in the larger one. Locate the line of penetration in

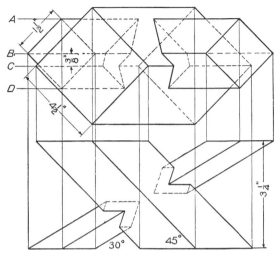

FIG. 115

the development of the smaller prism. A, B, C, and D are planes passed parallel to the vertical plane. Find the projections of each element cut from both prisms. Where they intersect determines the line of penetration, for each cut element lies in the same auxiliary plane. Number or letter each point. When objects are oblique to H or V, pass a plane to determine the true perimeter of the right section. The trace of such a plane in this problem must be perpendicular to the lateral edges of either prism. The development must be made from this sectional line and in a similar manner to the layout of the hexagonal prism in Exercise 1, Fig. 94 (page 95).

EXERCISE 3

As in Exercises 1 and 2, find the line of penetration of a right cylinder with a right cone (Fig. 116). Pass horizontal planes. Axes of both figures lie in the same plane. Use appropriate dimensions. Note that each plane cuts a circle from the cone and two elements from the cylinder. This is an illustration of a conical hopper connecting to a cylindrical pipe—or the gutter drip and rainwater pipe that can be seen on many houses.

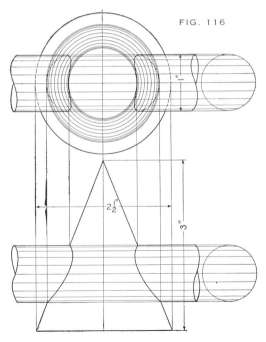

FIG. 116

$2\frac{1}{2}''$

$3''$

EXERCISE 4

A vertical pyramid is 4" high with a triangular base, with one edge of base making 15° with V; the length of one side is $2\frac{1}{2}$". It is penetrated by a horizontal 4" equilateral triangular prism with a perimeter of $6\frac{3}{4}$". One face is parallel to V and 1" from the vertical axis of the pyramid. The axis of the prism is $1\frac{3}{4}$" above the base. Draw three views full size, find the line of penetration, and develop both objects (Figs. 117, 118, and 119).

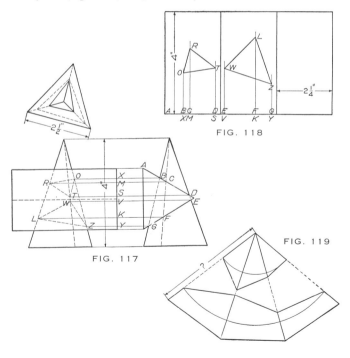

FIG. 118

FIG. 117

FIG. 119

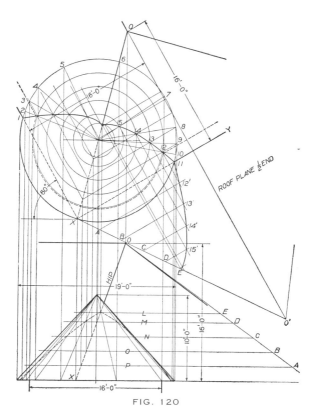

FIG. 120

EXERCISE 5

The conical steeple of a cylindrical tower is penetrated by the roof planes of a hip roof (Fig. 120). Find the line of penetration and lay out the developments of the conical roof and of the roof plane adjacent to the hip, showing the lines of penetration therein. Scale: $\frac{3}{4}$" = 1' 0".

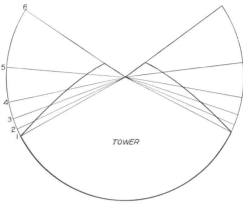

FIG. 121

After drawing the plan and elevation from Fig. 120, pass planes M, N, O, and P to determine the line of penetration. Because each plane is parallel to the base of the cone, it will cut a true circle from the cone, as shown in the plan. It will also cut a line from each roof plane parallel to the base of the hip roof. There will be a point of penetration wherever this element crosses the circle cut by the same plane. A series of similarly acquired points will determine the line of intersection.

To develop a roof plane, use the triangle method to revolve point O of the upper corner of the hip into the same plane as the base of the cone and the roof. O moves in a plane perpendicular to the axis XY. Points 10, 11, and 12 move perpendicularly to the roof lines drawn through A, B, and C. The development of the cone has been described (Fig. 121).

EXERCISE 6

Develop the pattern for the base of a fan from dimensions given in Figs. 122 and 123. The right and left sides of this base are elliptic cylinders, that is, they are not circular in cross section. The true size of the cross section cut by plane Tt' is shown at X in the plan. This is the line of development Tt' (Fig. 123). The lengths of each element can easily be laid out and the triangular faces added. Draw to suitable scale.

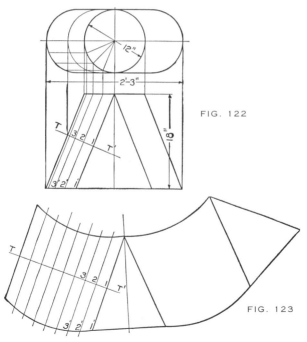

FIG. 122

FIG. 123

EXERCISE 7

Develop the slope sheet of a locomotive as given in Fig.
124, using triangulation for half of the drawing. This is
one of several practical problems to be derived from a
study of the locomotive for purposes of developments.

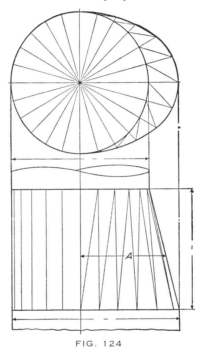

FIG. 124

The smokestack is another illustration of right cylinders
penetrating the outside cover of the boiler and requiring
templates or patterns.

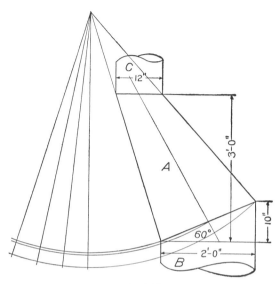

FIG. 125

EXERCISE 8

Draw the front elevation of the transition piece and develop by triangles (Fig. 125).

EXERCISE 9

The front face of a regular vertical triangular prism with a $10\frac{1}{2}''$ perimeter of and a 4" altitude is inclined backward and to the left at a 15° angle. A right square prism with a $6\frac{1}{2}''$ perimeter and a 4" altitude penetrates the former. Axes of both solids intersect at their center points. Develop both objects.

CHAPTER X

THE ISOMETRIC WORKING DRAWING

An isometric drawing is a pictorial or perspective representation. It is eminently useful to the artisan for clarifying hidden constructions. Among draftsmen it has supplemented the freehand perspective sketch because of the comparative ease with which it can be made.

The few principles of isometric drawing can be briefly summed up as follows:

a. All vertical edges in the object are vertical in the drawing, as in freehand.

b. All horizontal edges, representing right angles orthographically, make 30° to the horizontal in the isometric construction.

c. Non-isometric lines of edges making

other than right angles must be marked off orthographically first and then transferred to the
isometric drawing. This distorts the true length of
nonisometric lines but does not mar the pictorial
effect.

d. Surfaces not lying in the same plane are established from center-isometric axes.

e. Isometric circles are drawn within isometric
squares of the same diameter as the given circle.
Elliptic or irregular curves are constructed flat,
then transferred (Fig. 126).

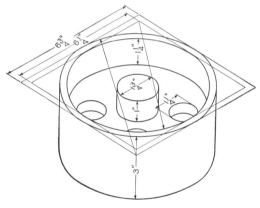

FIG. 126

f. Isometric workshop drawings are dimensioned.
Dimensions must be placed parallel to the isometric lines (Fig. 127).

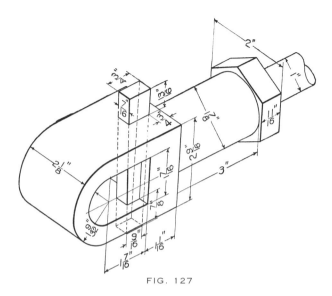

FIG. 127

g. Isometric drawings are usually shaded to accent the edges separating light from dark surfaces; the light is assumed to come from the left at a 45° angle. A better method, however—and one that enhances the pictorial effect—is to shade the edges nearest the observer's eye. This tends to lift the drawing of the object from the paper and relieve the unnatural effect of the isometric construction. Fig. 127 is an illustration of the stub end of a connecting rod and exemplifies the latter shading method described. The purpose of shading is to make the drawing more attractive, but aside from this it has no value, and many draftsmen do not shade their drawings at all.

h. Invisible lines are seldom shown in isometric drawings except when irregular lines are hidden by regular surfaces and the desired information can be shown in no other way.

The illustrations in this text are largely unshaded isometric drawings.

EXERCISE 1

Make an isometric drawing of a cardboard box or cigar box with the lid open. Scale: half size. No dimensions.

EXERCISE 2

Select a good-sized spool. Make an isometric drawing of it. Scale: double size. No dimensions.

EXERCISE 3

Copy the exercise of the connecting rod in Fig. 127. Scale: full size. Dimension.

EXERCISE 4

Figure 128 (page 128) represents the base and cap of a pattern for a pillow-block bearing. Scale: full size. Dimension.

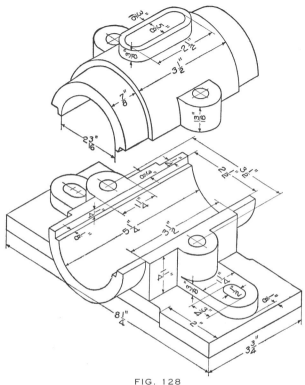

FIG. 128

EXERCISE 5

Make an isometric drawing of the teacher's desk to suitable scale. Do not show invisible lines. Dimension. Now substitute a bookcase for the desk.

EXERCISE 6

Make an isometric drawing of a mission chair. Look for non-isometric lines. Draw to suitable scale and dimension.

EXERCISE 7

Make an isometric drawing of a shaft-hanger (Fig. 129). Scale: half size.

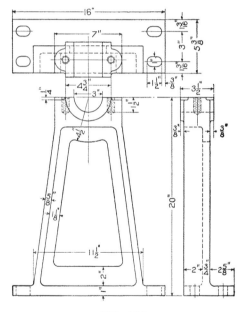

FIG. 129

CHAPTER XI

MISCELLANEOUS EXERCISES

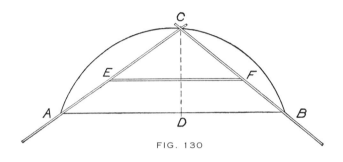

FIG. 130

EXERCISE 1

Construct the arc of a circle mechanically when it is inconvenient to determine its radius (Fig. 130). *Construction:* Make AB the chord of the arc ACB, keeping DC and ACB constant and the position changed so that points A and B remain in contact with lines AC and BC. The resultant points will determine the center and circumference of the required circle. The same problem might be constructed if two-by-fours are nailed together as the lines AC and BC suggest, with a third strip crossing lines AC and BC parallel to AB anywhere to hold the angle firmly in place.

EXERCISE 2

A graphical method for finding the distance AB across a pond when the land in triangle FED is inaccessible.

Construction: Set a stake at C in line with AB prolonged. Set another, D, so that C and B can be seen from it. Set a third stake, E, in line with BD prolonged so that DE equals BD. Set a fourth stake, F, at the intersection of EA and CD. Measure AC, AF, and FE. Show that AB is a fourth proportional to AF, AC, and (FE—AF). Draw a line through D parallel to AB. D bisects BE. DX is always $\frac{AB}{2}$ (Fig. 131).

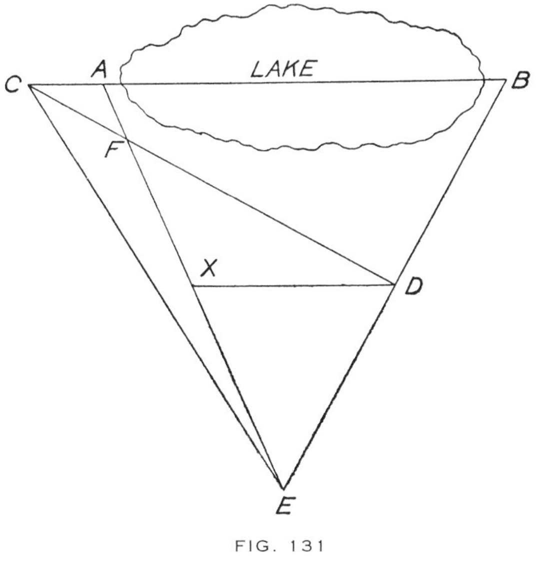

FIG. 131

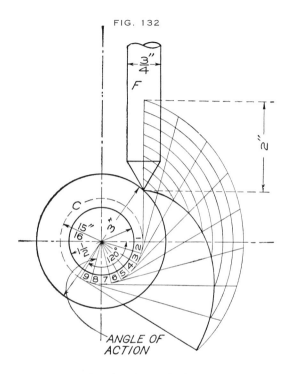

FIG. 132

$\frac{3''}{4}$

F

2"

C

$\frac{15''}{16}$ $\frac{3''}{-}$

$-\frac{1}{2}$ 20°

9 8 7 6 5 4 3 2

ANGLE OF
ACTION

EXERCISE 3

Draw an involute cam that involves the construction of the involute curve in Fig. 50 on page 54.

Construction: A cam is a very useful mechanical device that gives various motions to machine parts at regular intervals of time. It is generally a flat disk, although it is can also be cylindrical. Harvesters, printing presses, sewing machines, looms, and steam-valve mechanisms employ a considerable number of cam constructions (Fig. 132). To draw an involute cam with a given rise in a given angle of action, use the following information:

Let A = rise of the follower or throw, and

X = the radius of the base circle C.

As in the figure, the angle of action is 120°.

a = $\frac{2}{3}$ X, $\frac{2}{3}$ being the ratio of the arc through which the cam works, to a semicircle or straight angle.

2 = 44/21X, assuming π = $3\frac{1}{7}$.

X, or the radius, = 2/44/21 = 2 x 21/44 = $\frac{42"}{44}$, or nearly 1".

Assume $\frac{15}{16}$ as the most convenient. Using this radius, draw the base circle C and construct tangents upon which to lay out the involute curve. The machine itself will determine the diameter of the disk. Mark off the rise of the follower on tangent 1 and divide this into as many parts as tangents have been constructed. With the center at O, draw concentric circles to corresponding tangents from the points on the axis of the follower (F).

THE HELIX

EXERCISE 4

Fig. 133 (page 134) shows the development of a helical curve as unwrapped from a cylinder.

Imagine the surface of the cylinder laid out on paper. If

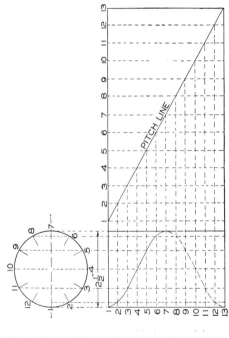

FIG. 133 DEVELOPMENT OF THE HELIX

a diagonal line were drawn and the paper wrapped about
the cylinder, the line would then illustrate the helix.

EXERCISE 5

The application of the helix can also be seen in coil
springs, two of which are illustrated in Fig. 134.

Construction: The constructions may be laid out as in
Fig. 50 (page 54). As in a screw thread, the pitch of the

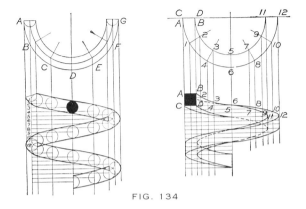

FIG. 134

helix is the distance between two opposite points lying on the curve and the same cylindrical element. In drawing the spring, use the helical curve as a centerline. Draw a number of small circles equal to the diameter of the round coil desired. The contour can easily be defined by drawing tangent helices to these circles. If square or rectangular material is used, draw the helices from each of the four corners, (A, B, C, and D) of the cross section.

EXERCISE 6

Make a 6-inch-long coil spring with an inside diameter of $3\frac{1}{2}$" and a $1\frac{1}{2}$" pitch from 5" of round steel.

EXERCISE 7

Make a 6 inch-long square spring with a $\frac{3}{4}$" pitch and a 4" outside diameter out of $\frac{1}{2}$" material.

EXERCISE 8

Determine the length of material required in each preceding exercise.

SHEET-METAL PROBLEMS

On page 95 references were made to the value of knowing how to lay out a pattern or template for sheet-metal problems; geometric methods in drafting are of especial use to sheet-metal draftsmen.

A great many problems dealing with sheet metal also involve warped surfaces. Such surfaces are cannot be developed by any regular method. Remember, for example, in the development of Fig. 97 (page 99), the cone was first divided into elements of regular intervals, say twelve in all, and their true length determined by revolving each foreshortened element parallel to the vertical view. Any two true elements laid out with the chord of their basal arc will form a triangle. Adjacent triangles are constructed in a similar manner and the pattern completed.

EXERCISE 9

Fig. 135 is an illustration of a transition piece for a smokestack or fan. Fig. 136 shows the pattern when laid out. Scale: 1" = 1' 0".

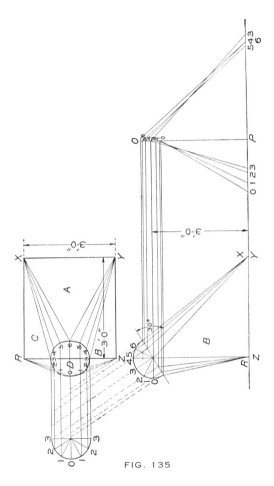

FIG. 135

Constructions: Planes A, B, C, and D are triangles whose true shapes can easily be determined from the projections. The four corners are sections of oblique cones that have been

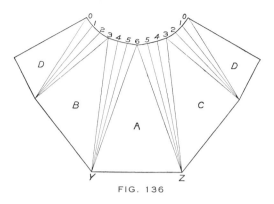

FIG. 136

previously described. A shorter method of measuring the true length of these elements is to find the hypotenuse of a right angle the base of which is the distance from X or Y to points 3, 4, 5, and 6 in the plan. The altitude of each triangle is the projected vertical altitude as seen in the front view. Mark off the side anywhere, as at OP, and draw each hypotenuse. These are the true lengths desired in the pattern between planes A, B, and C. The true lengths of other elements are found in a similar way. As the section of the top is a circle taken at an angle 30° from the horizontal, an auxiliary view will show a true circle, as in the front or top view. A semicircle will suffice. Divide it into an equal number of points for convenience, and project back to the corresponding plan and elevation. Connect these points with R, X, Y, and Z, and proceed with the development.

Mark off plane A first (Fig. 136). With 6 as a center, strike an arc equal to 6—5 (on the section) and the pattern with Y as a center and a radius equal to 5—5 at OP. Continue this process until each of the longer diagonals are used, half on each side of plane A. When all of the longer diagonals are used add planes B and C, and then add the diagonals laid out on the left of OP. The plane D is bisected to show a symmetrical development.

EXERCISE 10

Fig. 137 is the layout and working drawing of the base of a smokestack. The top of the base is circular while the ends are semioblique cones.

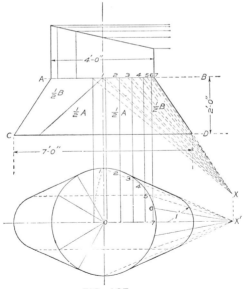

FIG. 137

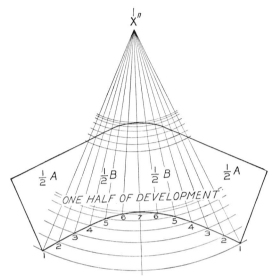

FIG. 138

Construction: Make a development of $\frac{1}{2}$ the lateral surface similar in shape to Fig. 136. Scale: 1" = 1' 0". To find the true length of all the elements, revolve the true length of everything parallel to the vertical plane upon which the front view is projected. Use X' as a center and X'—6 as a radius. Strike an arc upon 0—X'. Project up to the plane AB. Connect the newly found point with X, and the line will now be seen in its true length, parallel to V.

When the plane CD cuts the new position, X'—6 will be the true length of the portion that constitutes the surface and can be marked off on X"—6 of the development (Fig. 138).

EXERCISE 11

Fig. 139 is an illustration of scoop for a grocer's scale. The development of A is a pattern of the portion of a cylinder that a tinsmith would be required to lay out for a template. *Construction:* Substitute suitable dimensions and draw it, beginning with centerline X-X. Divide the end view of section A into a convenient number of similar, equal parts. Project back to X-X. Next, lay out the circumference of section A in development and project the points to corresponding places.

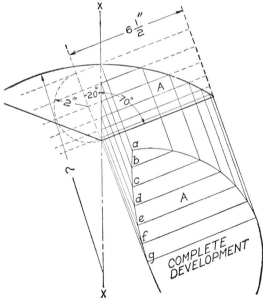

FIG. 139

SECTIONS OF
A WORKING
DRAWING

The ability to make sections in a drawing wherever possible constructive difficulties may later arise is a part of the value of what a draftsman does. Sections are very helpful for showing interior constructions; in a complicated drawing, they are an absolute necessity. In the illustrations, the sections are shown by crosshatched lines; the relationship of adjacent parts is shown by drawing the crosshatches at different angles. To determine the location of sections, pass planes through both the vertical and horizontal centers of the object. On pages 94, 95, and 96 and elsewhere in Chapter VIII, sections were made of geometric solids and their developments were required. Figs. 140 and 141 are sections of small machine parts and illustrate the practical value of such a construction. Fig. 142 is a sole plate for a pillow block. Make freehand sketches of each of the sectioned illustrations in order to get a pictorial view of the objects.

EXERCISE 12

An illustration of a core box for a pipe tee (Fig. 68) appears on page 66. Fig. 143 is an isometric illustration of a pipe-tee pattern but not for the box just referred to. Make three views from the illustration. Scale: full size.

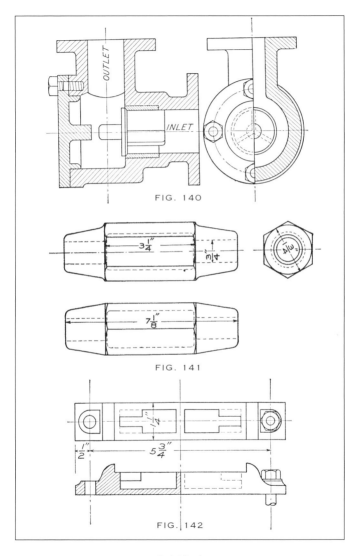

OUTLET

INLET

FIG. 140

$3\frac{1}{4}''$

$\frac{3}{4}''$

$\frac{3}{8}$

$7\frac{1}{8}''$

FIG. 141

$\frac{1}{4}''$

$\frac{1}{2}''$

$5\frac{3}{4}''$

FIG. 142

/ 143 /

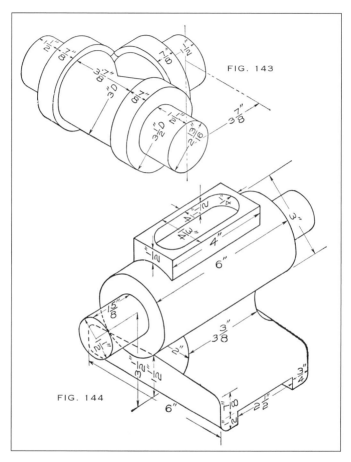

FIG. 143

FIG. 144

EXERCISE 13

An illustration of a pattern for a pedestal bearing appears in Fig 144. Make three views. Scale: 9" = 1' 0".

EXERCISE 14

An assembly drawing of a simple machine vise appears in
Fig. 145. Make a detail drawing, filling in those dimensions

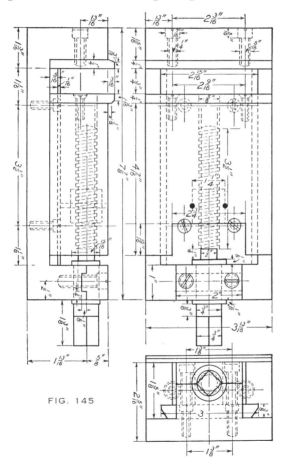

FIG. 145

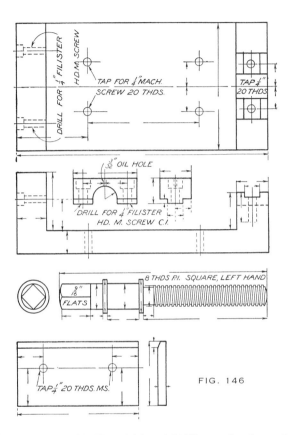

FIG. 146

that are omitted in Figs. 146 and 147 but only after refer-
encing the assembly drawing. Make a full-size drawing
of the assembled vise before the detail drawing. Make a
stock list as suggested on page 72. Look for any necessary alterations.

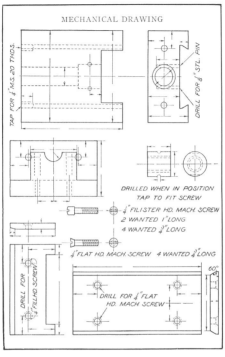

FIG. 147

The notes on the detail drawing make reference to the machinist's finishing the casting after it has been molded from the pattern. Such information is essential; it eliminates hazardous guesses and mistakes and saves time and material. The arrangement of pieces and parts should be very carefully planned on the drawing paper. Much more can then be placed on one sheet.

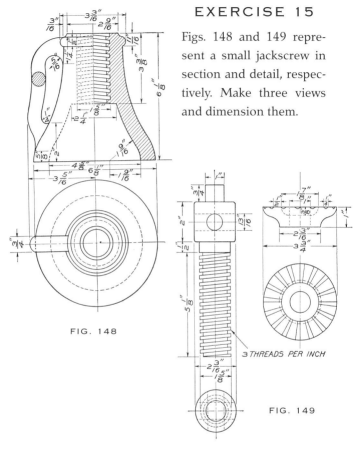

EXERCISE 15

Figs. 148 and 149 represent a small jackscrew in section and detail, respectively. Make three views and dimension them.

FIG. 148

3 THREADS PER INCH

FIG. 149

EXERCISE 16

A universal joint (also called a Hooke's joint) is illustrated in Fig. 150. Make three views and section.

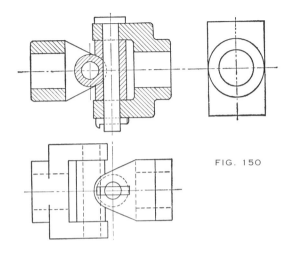

FIG. 150

EXERCISE 17

A side crank arm is illustrated in Fig. 151. Make three views and section at X-x.

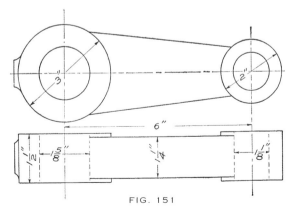

FIG. 151

EXERCISE 18

A turnbuckle is shown in Fig. 141 (page 143). Make three views and section.

Make a drawing of each preceeding exercise to scale as suggested. Use suitable diameters in each case and section.

EXERCISE 19

Make an isometric drawing of the mission footstool as shown in Fig. 152.

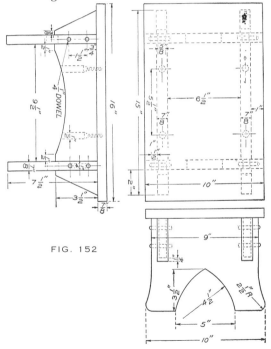

FIG. 152

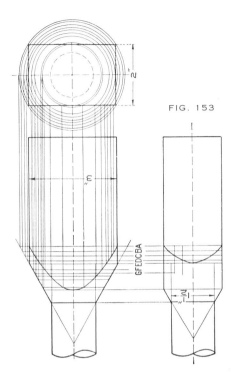

FIG. 153

INTERSECTIONS
AND PENETRATIONS

EXERCISE 20

The three views of the stub end of a connecting rod are shown in Fig. 153. To find the curve of intersection, cone, and prism, pass vertical planes A, B, C, and D, cutting both the cone and the rectangular prism. Each plane cuts a circle from the cone and a rectangle from the prism. Where these figures intersect is a point of the curve desired. The same method is used in finding a curve of intersection of a cone and hexagonal prism, or the chamfered portion of a hexagonal nut.

LETTERING EXERCISES

Those who have a great deal of difficulty with lettering often have only a vague idea of the shape of each individual letter. No draftsman, however experienced, can produce well-formed letters without a clear picture of each letter's shape. The beginning student should read and follow these suggestions closely. Use practice paper before commencing one of the exercises below.

INSTRUCTIONS

1. Make the vertical stroke of the letter A first, then the slanting stroke.

2. Make the bottom part of the letter B wider than the upper.

3. The letters C, G, and Q are modifications of the letter O.

4. Keep the bottom part of the letter D full.

5. The lowest bar of the letter E is a little longer than the upper bar. The middle bar is the shortest and slightly above the center, as in an F.

6. Draw the two outside bars of the letter H first. The horizontal bar is slightly above center.

7. The letter J is a portion of the letter U.

8. Make the short bar of the letter K slope from the upper end of the first bar.

9. The letter M is broad. First draw the two outside bars parallel, then the intermediate strokes. Draw the letter N in the same manner.

10. The letters P and R are similar. Keep the tops full.

11. The letter S is best made inside the letter O, with the bottom part a little wider and fuller.

12. Make the first bar of the letter W slope slightly to the right. Keep the letter broad.

13. The letter V is the letter A upside down.

14. The widths of all letters should always be kept in proportion to their heights.

15. Draftsmen commonly employ sloping Gothic letters. Vertical letters are sometimes used, but are more tedious to make look good.

EXERCISE 21

Draw Fig. 154 on $\frac{1}{8}$" graph paper. Each small figure pertains to the number of boxes on the graph paper.

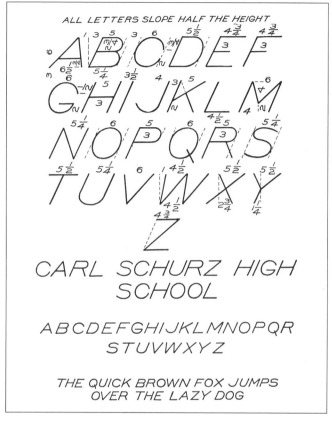

FIG. 154

This exercise is a study in form and proportion and should be executed with precision and care. Use a 2H pencil with a conical point.

```
    PENCIL EACH LETTERING SHEET ON
CROSS SECTION PAPER PROVIDED FOR THAT
PURPOSE AND SUBMIT EACH PENCILLED
LINE TO THE INSTRUCTOR FOR HIS
CRITICISM.   IN DOING THIS THE STUDENT
WILL SAVE TIME AND IMPROVE HIS
STANDARD OF WORKMANSHIP MORE RAPIDLY.
    USE A 2H PENCIL CONICAL POINT.   ALL
DRAWINGS SHOULD BE KEPT NEAT AND
CLEAN.  SPACES BETWEEN WORDS SHALL
NOT VARY FROM LESS THAN 2 TO MORE
THAN 3 DIVISIONS ON THE CROSS SECTION
PAPER.   SPACES BETWEEN PARAGRAPHS
SHOULD BE DOUBLE THE SPACE BETWEEN
LINES.   A SINGLE SPACE IS SUFFICIENT
TO SEPARATE LINES IN THE PARAGRAPH.

    INDENT EACH NEW PARAGRAPH.   USE A
LARGE PEN HOLDER WITH A 5I6 EF PEN
POINT.   KEEP THE PENPOINT CLEAN TO
ALLOW A STEADY FLOW OF INK.   THE INK
CLOGS THE PEN'S ACTION VERY QUICKLY.
DO NOT FILL THE PEN FULL OF INK IF YOU
WISH TO DO EVEN LETTERING AND AVOID
BLOTS.

    INDENT


        REPEAT LAST PARAGRAPH
```

FIG. 155

EXERCISE 22

Fig. 155 is an exercise in lettering one space high. It is an application of Exercise 21.

EXERCISE 23

Fig. 156 is an exercise of figures and fractions on graph paper. After reading the directions in Fig. 155, fill in all vacant spaces, striving for uniformity as in previous

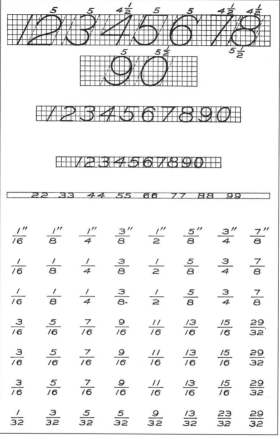

FIG. 156

exercises. Hard practice is a good master. The draftsman's figures are quite different from printed figures and
should therefore be scrutinized closely.

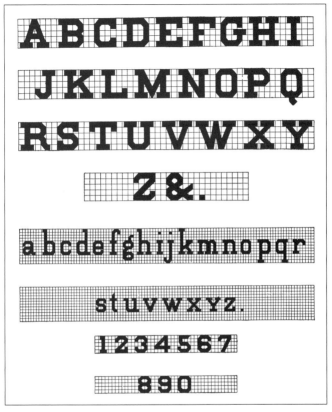

FIG. 157

EXERCISE 24

Fig. 157 is an exercise in block lettering quite often used in designs for covers, titles, headings, etc. Note that there are five divisions of the height instead of eight.

CHAPTER XII

A SUGGESTED COURSE FOR YOUNG DRAFTSMEN

Those who expect to study design are not required to complete the entire course of mechanical drawing.

GROUP 1: GEOMETRIC EXERCISES

EXERCISE 1

Bisect a given right line and arc.

EXERCISE 2

Erect perpendiculars to a given line using any method.

EXERCISE 3

Draw parallel lines using two different methods.

EXERCISE 4

Divide a given line into proportional parts.

EXERCISE 5

Construct tangents to a given arc of any radius.

EXERCISE 6

Duplicate and bisect a given angle.

EXERCISE 7

Without triangles, construct angles of 30°, 60°, 75°, 45°, 22°-30', and 37°-30'.

EXERCISE 8

Using only triangles, divide a semicircle into angles of 15°.

EXERCISE 9

Rectify a quadrant of a circle using two different methods). Approximate.

EXERCISE 10

Triangles (trillium, trefoil).

Draw each of the following:

   a. A right triangle (rise, run, and pitch of a gable roof rafter)

The following formula can be used to find the distance across an unknown area, a stream, lake, or park, or the altitude of a tree:

$$x^2 + y^2 = z^2$$

   b. An equilateral (isometric square, Gothic arch)

   c. An isosceles

   d. A scalene

Queries: How can the area of any triangle be found? What is the sum of all angles of any triangle?

EXERCISE 11

Draw a square (bolt head plan, syringa).

EXERCISE 12

Polygons (pansy, violet, crystals).

Draw each of the following:

   a. A pentagonal star (using three different methods)

   b. A hexagonal bolt head plan (using two methods)

   c. A heptagon

   d. An octagonal taboret top

   e. A combination of a, b, c, and d on a given side of 1".

Prove all polygons by the formula $\dfrac{2n - 4 \times 90°}{n}$ when

n = the number of sides of the polygon. Use a protractor to verify the results.

EXERCISE 13

Circles.

The problems marked by an asterisk may be omitted.

Draw each of the following:

  a. Three circles within an equilateral triangle

  b. Circles tangent to each other and the given circle, inside or outside it

  c. A Gothic arch

  d. A circle tangent to a given circle and line

  e. A circle tangent to two given circles that are not tangent to each other

  Note: The smallest circle is not acceptable.

*f. A shaft $1\frac{1}{2}$" in diameter rotates within a ball bearing consisting of twelve tempered steel balls showing the size of the balls required

  g. Four circles within a square.

  h. A Maltese cross

  i. Geometric circular borders

  j. Cavetto, cyma, reversa, cyma recta, ogee, and scotia moldings

EXERCISE 14

Ellipses and elliptic curves (ecliptic conic sections).

The problems marked by an asterisk may be omitted.

  a. Draw an ellipse using the focal, the circle, and trammel methods.

Now draw each of the following:

b. A five-pointed elliptic arch

c. Greek, Persian, and Gothic arches

d. An elliptic cam

*e. The path of a point on a connecting rod in one revolution

*f. A cycloid (gear teeth)

*g. An epicycloid (gear teeth)

*h. A hypocycloid (gear teeth)

*i. A parabola (conic sections)

*j. A hyperbola (conic sections)

EXERCISE 15

Spirals.

The problems marked by an asterisk may be omitted.

Draw each of the following:

a. An Archimedean spiral of one or more whorls

b. An Ionic volute (Ionic capital)

*c. A heart plate cam (sewing machine, bobbin winder)

*d. An involute (gear teeth)

*e. A helix (screw, thread, clutch coupling)

BIOGRAPHICAL

Read the biographies of Archimedes, Pythagoras, Euclid, and Giacomo Barozzi da Vignola in any encyclopedia.

GROUP 2: PROJECTIONS

I. WORKING DRAWINGS

Draw each of the following:

1. Three views of a cylinder

2. Three views of a prism

3. Two views of a plinth, one view given

4. Three views of a pyramid

5. A hexagonal nut

6. A crank arm

7. A small pedestal bearing

   Note: Do not dimension the previous seven problems. Substitutes may be selected for these objects if and when they are not available or advisable. Exercises 8 to 15 are to be dimensioned carefully. Use models rather than drawings at the beginning of this course so that the absolute relation of object to drawing is established as early as possible.

8. A taboret or stand

9. A coat hanger

10. A cardboard box

11. Any part of a lathe

12. A work bench

13. Detailed working drawings from machine parts

14. Working drawings from isometric blueprints

15. Freehand working drawings from sketches (freehand)

II. REVOLUTION
AXES OF SYMMETRY

1. Draw three views of a prism, plinth, or pyramid.

2. Draw three views of No. 1 when revolved about a vertical axis 30° counterclockwise.

3. Revolve the object from No. 2 about side axis through 30° to the left.

4. Revolve the object from No. 1 forward 20° about a front axis.

5. Revolve the object from No. 2 backward 25° about a front axis.

6. Revolve the object from No. 5 to the right 30° about a side axis.

7. Revolve the object from No. 4 about a vertical axis 15°.

8. Revolve the object from No. 5 to the right 15° about a side axis.

Several plates involving the modified positions of geometric figures should be drawn so the theory of projections and the three laws of revolution (Chapter VII) can be learned.

III. THE POINT, LINE, AND PLANE

(For advanced students.) Draw in both first and third angles.

1. Find H and V projections of a point a) $1\frac{1}{2}$" in front of V and $2\frac{1}{2}$" above H; and b) 2" below H and $1\frac{1}{2}$" behind V. Always open the first angle.

2. Draw the projections of a line which is parallel to the H and V planes, $1\frac{1}{2}$" above H and 2" in front of V.

3. Draw two views of a line a) oblique to H and parallel to V; b) oblique to V and parallel to H; and c) oblique to H and V.

4. Find the true length of the lines in No. 3. What is the difference between the projected length and the true length of a line?

5. Pass a plane a) parallel to H; b) parallel to V; c) parallel to P; d) perpendicular to H and any acute with V; e) perpendicular to V and any acute with H and P.

6. Find the intersection of a and b, and also c and d, in No. 5.

IV. THE DEVELOPMENT OF SURFACES FOR PATTERNS

1. Parallel lines. Cylinders, prisms, etc.

2. Radial lines. Cones, pyramids, etc.

3. Method of triangles. Warped surfaces.

4. Method of revolution. Frustums and truncations.

5. Method of parallel planes, oblique planes. Penetrations.

GLOSSARY

For the benefit of those who meet terms and expressions in this manual for the first time, the following vocabulary, with definitions, is provided:

Altitude: Vertical height.

Angle: Space between two intersecting lines.

Apex: Point where converging lines meet.

Arc: Any part of the circumference of a circle.

Area: Surface in units of measurement.

Bisect: To cut in two equal parts.

Bisector: A line that bisects.

Chord: The line connecting any two points of an arc of a circle.

Circumference: The boundary of a circle.

Circumscribe: To draw around.

Convention: The customary method or symbol used in producing a drawing.

Decagon: A figure with ten sides and ten angles.

Degree: $\frac{1}{360}$ of a circle.

Diameter: The distance measured across the center of a circle, or a line drawn through the center terminating in the circumference.

Element: A part that goes to make up the whole.

Elevation: A view of an object looking at the front or side.

Ellipse: An oval.

Elliptic: Pertaining to the shape of an ellipse.

Equilateral: Having equal sides.

Frustum: The remaining portion of a cone or pyramid when the top has been removed parallel to its base.

Hemisphere: Half a sphere.

Heptagon: A figure of seven sides and seven angles.

Hexagon: A figure of six sides and six angles.

Horizontal: Parallel to the horizon.

Hypotenuse: The diagonal distance between opposite angles of a rectangle, or the side opposite the right angle.

Involute: Curved or spiraled or inward.

Isometric: Of equal measurement.

Isosceles triangle: A triangle with two sides of equal length and equal base angles.

Lateral: Side.

Line: That which has length only.

Median: A line drawn from the vertex of an angle to the middle point of the opposite side of a triangle.

Mensuration: The application of geometry to the computation of lengths, areas, or, especially, volumes.

Nonagon: A figure of nine sides and nine angles.

Octagon: A figure of eight sides and eight angles.

Orthographic: Derived from two Greek words, *orthos,* "straight," and *graph,* "to write," it is used to describe a straight-line drawing that is projected horizontally, vertically, and in profile as if all three views were in the same plane. In an orthographic projection, the lines appear perpendicular to the drawing surface.

Parallel: Lines or planes are parallel when all points of one are equally distant from all points of another.

Parallelogram: A four-sided figure with opposite sides parallel and of equal length.

Pentagon: A figure of five sides and five angles.

Perimeter: The distance measured around an object.

Perpendicular: Any line at right angles to another.

Perspective: A two-dimensional drawing that represents the spatial relation of objects as they appear in three-dimensions through the use of converging parallel lines, thereby creating the illusion of depth and distance.

Pi (π): A Greek letter used as a symbol to express the relation between the diameter and circumference of a circle. If π = 3.1416, the diameter of a circle multiplied by π equals its circumference.

Plan: A view looking down upon the top.

Plane: A surface with length and width but no thickness.

Plinth: A prism whose height is less than any one of its other dimensions.

Point: That which has position only.

Polygon: A closed figure bounded by straight lines.

Prism: A figure bounded by rectangular faces, two of which are parallel.

Project: To point toward.

Pyramid: A solid with triangular faces converging to a common vertex.

Quadrant: The fourth part of a circle.

Quadrilateral: A four-sided polygon.

Radius: Half the diameter of a circle.

Radii: The plural form of radius.

Rectangle: A closed figure with four right angles of 90° each.

Rectify: To make straight or right.

Rectilinear: Pertaining to right or straight lines.

Rotate: To roll.

Scalene triangle: A triangle all of whose sides are unequal in length.

Section: A view determined by a cutting plane.

Sector: A radial division of a circle, or the space between two radial elements.

Segment: The space between the chord and arc of a circle.

Semicircle: Half a circle.

Sphere: Also called a ball or globe, a solid whose surface points are all equidistant from a point within called the center.

Tangent: Lying adjacent at a single point.

Triangle: A three-sided figure.

Trisect: To cut into three equal parts.

Truncate: To cut off.

Vertex: A common point of several converging lines.

Vertical: Straight up and down.

DEFINITIONS
OF SYMBOLS

$2\pi r$: Circumference of a circle when r = radius

πr^2: Area of a circle when r = radius

⊥　　Perpendicular

∥　　Parallel

=　　Equals or is equal to

∢　　Angles

×　　Intersecting, or multiplied by, as the case may be

∴　　Therefore

∟　　Right angle; two intersecting lines at 90° to each other

∠　　Acute angle; two intersecting lines less than 90° to each other

╲＿　Obtuse angle; two intersecting lines more than 90° to each other

CP:　Circular pitch

DP:　Diametral pitch

H:　　Horizontal

V:　　Vertical

P:　　Profile

H:　　Height

W:　　Width

D:　　Depth

r:　　Radius

D:　　Diameter

GL:　Ground line

VL:　Vertical line

INDEX